BASIC LATIN

FOR PLANT TAXONOMISTS

BY

A. BARANOV

(Cambridge, Mass., USA)

Authorized Reprint

3301 LEHRE

VERLAG VON J. CRAMER

1971

Sole Agents for the USA
Stechert-Hafner Service Agency, Inc., New York

Sole Agents for the British Commonwealth
Wheldon & Wesley, Ltd., Codicote, Herts.

Authorized reprint
of the New Dehli 1968 edition
(separately printed from vol. 21 of "Advancing Frontiers
of Plant Science")
Printed in Germany 1971
by Strauss & Cramer GmbH, 6901 Leutershausen
ISBN 3 7682 0727 7

Foreword to second printing

For many years I was strongly convinced that there is a definite need for such a book as my 'Basic Latin for Plant Taxonomists'. Then, before long my conviction was confirmed for the stock of the first printing in 1968, limited to 150 copies, within a few months became depleted.

Therefore, it is most gratifying for me to have the timely interest of J. Cramer in reprinting this book. Through their efforts not just another rare natural history publication will become available, but rather a basic and many-purpose tool. This is because the book may serve equally good as a text for the students, as a plain and sound basis of a course in botanical Latin for the professors, and as an useful working guide for literally thousands of botanists who in this modern age lack basic training in the classical languages and for this reason are perplexed by botanical Latin.

Correction

The reprinting of this book has offered an opportunity to make one important correction in its text which is as follows.

On the page 81 [item (e)] it is said that the Latin noun *verticillaster* might be declined in two ways. However, this is not so. Actually this noun belongs to the second declension of masculine Latin nouns and follows the pattern of declension of such Latin words as *vir*, *puer*, etc.

In view of this the noun *verticillaster* must be excluded from the third declension of masculine nouns ending in -er [p. 81, item (e)] and transferred to the second declension of masculine nouns ending in -ir/-er [p. 77, item 2].

Those interested in etymology of the noun *verticillaster* and other details about the grammatical nature of this word have to see a special note by this writer: "Apropos of the Word "Verticillaster"" published in TAXON 18(4):429-430, August, 1969.

CONTENTS

SECT. I. THE NATURE AND METHODS OF BUILDING UP PLANT NAMES, KEYS AND DIAGNOSES

FOREWORD

The writing of this book was prompted by the realization that plant taxonomists must write formal descriptions of new taxa in Latin; but at the time when compilation of this guide was started, in the early 1950's, there was no book available instructing in the preparation of these descriptions, such knowledge being handed down mostly by tradition.

Therefore, in case of any difficulties with the usage of botanical Latin which may confront the taxonomist, the latter had been usually compelled to consult standard grammar texts which are mostly not helpful in this instance, as the writer knows by his own experience. For such purposes, one requires a special book in which the vocabulary, the examples illustrating rules of grammar, etc., have a special botanical character.

Presently, there are available several manuals of botanical Latin, but this book is different from all of them in that the former do not instruct directly *how* to write a plant description and give only suggestions. This guide, which is an attempt to contribute to this need, is written from a practical point of view. The emphasis in it is on the methods of writing plant descriptions, especially the descriptions of species. The theoretical part of the book is simplified and made as succint as possible, and is totally adapted to the main purpose of the book.

The basis for this guide was supplied by the course of Latin botanical terminology given by me to the junior research fellows of the Institute of Forestry of Academia Sinica in 1951 and in 1958.

A word here on how to use the book. It is divided into five sections, the methods of plant description being concentrated into sections I-II, while all theoretical explanations about the principles and facts of Latin grammar and botaincal usage of this language

2

are contered in sections III-V. Therefore, in case of any difficulties in understanding the first two sections, the readers have to refer for explanations to the last sections.

I hope that this book, intended primarily for students and beginners in plant taxonomy, will meet the need of those having difficulties with preparing Latin plant descriptions.

As a final note, I wish to thank my colleagues at the Botanical Museum of the Harvard University and at the Harvard University Herbaria for their willingness to discuss the problems cencerned with the preparation of this guide. I am especially grateful to Dr. L. A. Garay, Dr. R. A. Howard, Dr. R. E. Schultes and Mr. C. Schweinfurth who helped me in various ways in this work.

SECTION I. THE NATURE AND METHODS OF BUILDING UP PLANT NAMES, KEYS AND DIAGNOSES.

1. The names of plant taxa.

§ 1. General remarks.

1. Scientific names of plant taxa are composed of Latin or latinized words taken from other languages. It is not the purpose of this work to deal with all kinds of such names. Therefore here will be given only basic information about the names of genera and species.

2. The choosing of the name for a new taxon is left to the discretion of its author. There are, however, some regulations relating to the choosing of names in the International Code of Nomenclature (Articles 20-23). These regulations must be consulted and observed.

§ 2. The plant names.

1. According to binomial system the name of each species consists of two words, the first word is the name of genus to which the plant belongs and the second word is the name of particular species in that genus. Therfore the *first* word is known as the *generic name* and the *second* one as the *specific name*. These two words together from a *binomial* or binary epithet.

§ 3. The generic name.

1. The name of a genus is a noun in Singular and it is always capitalized. Generic name may be taken from whatever source. For instance, it may be :

a. Ancient classical Latin name for the plant.

  Examples : Quercus [oak tree], Fagus [Beech tree], Rosa [ros

b. Name given in honor of some person.

Examples: Prinsepia [from James Prinsep], Adansonia [from Adanson], Kochia [from Koch], Linnaea [from Linnaeus].

c. Name formed artificially of Greek and/or Latin words describing some character typical for the members of this genus.

Examples: *Sagittaria* [Latin, after the leaf form of the species of this genus]; *Trisetum* [Latin, three bristles, the florets in this genus are three-awned]; *Pentstemon* [Greek, five and stamens; after five stamens in the flower]; *Leucophyllum* [Greek, white and leaf; after white foliage].

d. Name formed from another generic name as its anagram.

Examples: *Anogra* [anagram of Onagra], *Ifloga* [anagram of Filago], *Restella* [anagram of Stellera], *Tapiscia* [anagram of Pistacia].

e. Name formed from another generic name with the help of certain suffixes or prefixes.

Examples: *Paracroton* [prefix para + croton], *Neomyrtus* [prefix neo + myrtus], *Semiaquilegia* [prefix semi + aquilegia], *Trapella* [Trapa + diminutive suffix-ella], *Valerianella* [Valeriana + diminutive suffix -ella].

f. Aboriginal or vernacular name of the plant. Such names taken from the languages other than Latin are latinized or considered as the Latin ones.

Examples: *Bixa* [South-American name of the plant], *Tsuga* [Japanese name of the plant], *Alhagi* [Mauretanian name of the plant].

2. Generic name can not consist of two separate words; they must be either united into one word or hyphenated.

Examples: *Crataegomespilus, Sorbaronia, Ottoschmidtia, Neves-Armondia.*

3. Below are explained some cases of formation of generic names from the names of persons and from other generic names.

Generic names derived from the personal names with the help of suffixes *-ia, -a, -iella,* or *-ella* are all feminine nouns (belong to the first declension) and are formed in the following way:

a. Suffix *-a* is added to the end of the name if the name ends in a vowel, except the letter *a*. In latter case suffix *-ea* is added.

   Examples: *Bunge-a* from Bunge, but *Barbara-ea* from Barbara [feminine name], *Olga-ea* from Olga [feminine name] and *Lisa-ea* from Lisa.

b. Suffix *-ia* is added if the name ends in a consonant, except the termination *-er*. In latter case suffix *-a* is added.

   Examples: *Bernard-ia* from Bernard, *Berg-ia* from Berg, but *Steller-a* from Steller and *Winkler-a* from Winkler.

c. Suffix *-a* is added also to latinized names. Before the adding of this suffix the termination *-us* in these names is taken away.

   Examples: *Caesalpini-a* from Caesalpinius, *Linnae-a* from Linnaeus, *Clusi-a* from Clusuius.

d. The suffix *-iella* is joined directly to the end of names if they end in a consonant, except the letter *-r*. In the latter case suffix *-ella* is added.

If the name ends in a vowel the latter is taken away and to remaining stem suffix *-iella* is added. The exception is again with the stems ending in *-r*. In this case suffix *-ella* is added.

   Examples: *Hoffmann-iella* from Hoffmann, but *Engler-ella* from Engler; *Fedtschenk-iella* from Fedtschenko, but *Matsumur-ella* from Matsumura.

4. New generic names may be formed from the old ones by adding to the latter:

a. Prefixes, iso-, macro-, meta-, micro-, neo-, para-, pseudo-, semi-. These prefixes are directly joined to the name.

   Examples: iso + Trema = Isotrema; macro + Zamia =

Macrozamia; meta + Sequoia = Metasequoia; micro + Biota = Microbiota; neo + Myrtus = Neomyrtus; para + Croton = Paracroton; pseudo + Panax = Pseudopanax; semi + Aquilegia=Semiaquilegia.

b. Suffixes, -aria, -ella, -opsis which are joined to the stem of the name.

Examples: Spergul(a) ⟶ Spergul + aria = Spergularia
Valerian(a) ⟶ Valerian + ella = Vallerianella
Rhamn(us) ⟶ Rhamn + ella = Rhamnella
Steller(a) ⟶ Steller + opsis = Stelleropsis
Coryl(us) ⟶ Coryl + opsis = Corylopsis

As shown in the examples in order to receive stem of the name it is necessary in the nouns of the first and second declensions to take away the terminations -a or -us respectively. In the nouns of third declension it is necessary to put the latter in Genitive Singular and then take away the termination -is and after this to join the suffix to the stem.

Example: Stachys, id(is) Stachyd + opsis = Stachydopsis.

5. For the names of subgenera or sections nouns or adjectives are preferably chosen which describe some typical morphological feature of the members of that subgenus or section. If the epithet is an adjective it is always put in Plural.

Examples: Subgenera,   Sections,
               Fragariastrum   Macrocalyx
               Ulmaria   Pes-gallinaceus
               Phaca   Verticillati
               Pes-leonis   Latifoliae

6. For the names of series, there are often taken specific names of the typical species of that series, the adjectives being in Plural.

Examples: Diversifolii   Grandiflorae
               Macranthi   Paniculatae

In certain cases, however, for the names of sections are chosen words that describe geographical range or ecology of the members of the series.

Examples :   Asiatici   Montanae
                Petreae   Palustres

§ 4.   The specific name.

1. Specific epithet, same as generic name, may be an adjective, a noun, or it may be a combination of adjective and noun. It may be taken from any source whatsoever.

    The specific epithet may be :

    a. Latinized aboriginal (vernacular) name of the plant.

    Examples : Acer *mono*, Polygonum *posumbu*, Caragana *chamlagu*.

    b. Name given in dedication to some person.

    Examples : Kochia *Sieversiana*, Populus *Davidiana*, Quercus *Fabri*, Prunus *Maackii*.

    c. Name derived from some geographical name.

    Examples : Physocarpus *amurensis*, Polygala *sibirica*, Prinsepia *sinensis*.

    d. One of the adjectives describing some trait typical for the given species.

    Examples : Potentilla *viscosa*, Polygonum *viviparum*, Salix *pentandra*, Artemisia *laciniata*.

2. a. Specific epithet usually consists of one word.

    Examples : Rosa *daurica*, Ulmus *pumila*, Aruncus *asiaticus*.

    b. It may consist also of two or more words. In this case these words must be united into one word or hyphenated.

    Examples : Solanum *dulcamara*, Eriocaulon *chinorossicum*, Echinochloa *crus-galli*, Ipomoea *pes-caprae*, Doxantha *unguis-cati*.

3. Specific epithets—adjectives must agree grammatically with generic name (cf. also p. 122,123 of this guide). For the proper agreement of generic and specific names it is necessary to know grammatical gender of generic name.

Examples of agreement of generic and specific names :

Masculine gender.

Astragal*us* dauric*us*, Melilot*us* alb*us*, Astragal*us* chinens*is*.

Feminine gender.

Ros*a* dauric*a*, Mor*us* alb*a*, Eclipt*a* alb*a*, Calaminth*a* chinens*is*.

Neuter gender.

Menispermum dauricum, Lamium album, Chrysanthemum chinense.

Since the elements of a binomial are grammatically agreed, the transfer of a species from one genus to another entails the change of the termination of the specific epithet in accordance with the gender of a genus in which the species is transferred.

Examples :

Panic*um* italic*um* when transferred to Setaria becomes Setari*a* italic*a*.

Aln*us* fruticos*a* „ „ „ Alnaster „ Alnaster fruticos*us*.

Polygon*um* folios*um* „ „ „ Persicaria „ Persicaria folios*a*.

Euphorbi*a* lunulat*a* „ „ „ Galarhoe*us* „ Galarhoe-us lunulat*us*.

4. Specific epithets, when they are nouns, do not necessarily agree grammatically with generic name. These epithe*s* may be :

a. Noun in apposition or noun in Nominative case.

Examples : Cyperus *papyrus*, Dioscorea *batatas*, Punica *granatum*, Ocimum *basilicum*.

b. Noun in Genitive case.

Examples : Scabiosa *Fischeri*, Alhagi *camelorum*, Artemisia *desertorum*, Artemisia *terrae-albae*, Colocasia *antiquorum*.

The epithets in this group are Genitive form of personal

names or names of peoples, animals, countries, etc. When they are not a personal name they are usually put in Plural.

5. Special group of specific epithets are those combined by two hyphenated words; these words may be :

a. Two adjectives.

Examples : Cleyera *albo-punctata*, Aconitum *albo-violaceum*, Polygonum *hastato-triblobum*.

b. A noun and adjective.

Examples : Calonyction *bona-nox*, Pedicularis *sceptrum-Carolinum*, Alisma *Plantago-aquatica*, Veronica *anagallis-aquatica*.

Such compound epithets must agree grammatically with generic name *only if the last word in the compound is an adjective*.

6. The formation of basic types of specific epithets, if adjectives, is made in a following way.

a. When they are derived from nouns—geographical names they are formed by adding to the roots [stems] of geographical names of suffixes :

-alis,   *-cus,*   *-ensis,*   *-eus,*   *-nus,*   *-ticus* for masculine gender

-alis,   *-ca,*   *-ensis,*   *-ea,*   *-na,*   *-tica* for feminine gender

-ale,   *-cum,*   *-e,*   *-eum,*   *-num,*   *-ticum* for neuter gender

Examples :

Noun	Adjective		
	masc.	fem.	neut.
Mongolia, -ae	mongoli-cus,	mongoli-ca,	mongoli-cum
Oriens, -entis	orient-alis,	orient-alis,	orient-ale
Pekinum,-i	pekin-ensis,	pekin-ensis,	pekin-ense
Europa,-ae	europa-eus,	europa-ea,	europa-eum
America, -ae	america-nus,	america-na,	america-num

Asia, -ae   asia-ticus,   asia-tica,   asia-ticum

These adjectives may be of first, second, or third declension.

7. In certain nouns it is possible to derive the adjectives only of first and second declensions ; in others only of third declension; in still others of all three declensions.

Examples :

Noun	Adjective	Declension
Europa, -ae	europaeus, -a, -um	only 1-st and 2-nd.
Asia, -ae	asiaticus, -a, -um	
oriens,-entis	orientalis, -e	only 3-rd.
Amur	amuricus, -a, um	All three declensions
	amurensis, -e	

8. The most commonly used specific epithets, derived from geographical names, are listed below in alphabetical order.

anglicus, -a, -um	English	japonensis,-is,-e	Japanese
arcticus, -a, -um	arctic	japonicus,-a,-um	,,
asiaticus, -a, -um	Asiatic	koreanus, -a, -um	Korean
australis,-is,-e	southern	(also coreanus, -a, -um)	,,
borealis,-is,-e	northern	koreensis,-is,-e	
brasilicus,-a,-um	Brazilian	(also coreensis,-is,-e)	Korean
brasiliensis,-is,-e	Brazilian	lusitanicus,-a,-um	of Portugal
britannicus,-a,-um	British	manshuricus,-a,-um	Manchurian
canadensis,-is,-e	Canadian	manshuriensis,-is,-e	Manchurian
chinensis,-is,-e	of China	mongolikus,-a,-um	Mongolian
danicus,-a,-um	of Denmark	norvegicus,-a,-um	Norvegian
		occidentalis,-is,-e	western
europaeus,-a,-um	European	orientalis,-is,-e	eastern
gallicus,-a,-um	French	polonicus,-a,-um	Polish
germanicus,-a,-um	German	rossicus,-a,-um	Russian
helveticus,-a,-um	Swiss	ruthenicus,-a,-um	Ruthenian
hispanicus,-a,-um	Spanish	septentrionalis,-is,-e	northern
indicus,-a,-um	of India	sibiricus,-a,-um	Siberian
italicus,-a,-um	Italian	sinensis,-is,-e	of China

sinicus,-a,-um	of China	suecicus,-a,-um	Swedish

9. Special groups of adjectives are those having endings : a. *-anus, -ana, -anum* and b. *-acus, -aca, -acum.*

Examples :

a.

africanus,-a,-um	African	bolivianus,-a,-um	Bolivian
alaskanus,-a,-um	Alaskan	chiloanus,-a,-um	Chilean
americanus,-a,-um	American	colombianus, -a,-um	Colombian
australianus,-a,-um	Australian	mexicanus,-a,-um	Mexican
brasilianus,-a,-um	Brazilian	peruvianus,-a,-um	Peruan

b.

aegyptiacus, a,-um	Egyptian	austriacus,-a,-um	Austrian
armeniacus,-a,-um	Armenian	syriacus,-a,-um	Syrian

10. Epithets of taxa (specific and infraspecific) derived from the personal names are formed by adding to these names of certain terminations.

a. If the name ends in *a consonant* the terminations are : *-ianus* for masculine, *-iana* for feminine and *-ianum* for neuter gender. But if the name ends in *-er* the terminations may be also, *-anus, -ana* and *-anum.*

Examples :

Name	Epithet (Adjective)		
	masc.	*fem.*	*neut.*
Komarov	Komarov-ianus	Komarov-iana	Komarov-ianum
Fischer	Fischer-ianus	Fischer-iana	Fischer-ianum
	or Fischer-anus	Fischer-ana	Fischer-anum

b. If the name ends in *-e, -i, -o, -u, -y* the terminations are : *-anus, -ana, -anum.*

Examples:

Name	Epithet (Adjective)		
	masc.	*fem.*	*neut.*
Linne	Linne-anus	Linne-ana	Linne-anum

Lipski	Lipski-anus	Lipski-ana	Lipski-anum
Karo	Karo-anus	Karo-ana	Karo-anum
Gray	Gray-anus	Gray-ana	Gray-anum

c. If the name ends in -*a* the terminations are : *-eanus, -eana, -eanum.*

Examples :

Name	Epithet (Adjective)		
	masc.	*fem.*	*neut.*
Noda	Noda-eanus	Noda-eana	Noda-eanum
Ivanova	Ivanova-eanus	Ivanova-eana	Ivanova-eanum

11. These epithets, as real adjectives, are declinable according to the first and second declensions, and must necessarily agree grammatically with generic names.

12. Often personal names in Genitive case Singular are used as specific or infraspecific epithets, in the sense and instead of adjectives. These epithets are also formed by adding to the names of certain terminations.

a. If the name ends in a *vowel* or in -*y*, to it is added letter -*i*. Important note : to the names ending in -*a* the letter -*i* is added *only* if these names are masculine.

Examples :

Name	Epithet
Karo	Karo-i
Gray	Gray-i
Bunge	Bunge-i
Hara	Hara-i
Noda	Noda-i

b. If the name ends in a *consonant* (except letters *er*) to it are added letters -*ii*.

Examples :

Name	Epithet
Komarov	Komarov-ii
Ledebour	Ledebour-ii

f the name ends in *-er*, letter *-i* is added to it.

Examples :

Name	Epithet
Fischer	Fischer -i
Hooker	Hooker -i

d. If the name is latinized it takes ending corresponding to its Genitive Singular form.

Examples :

Name	Epithet
Linnaeus	Linnae-i
Clusius	Clusi-i
Theophrastus	Theophrast-*i*
Avicenna	Avicenn-*ae*
Victor	Victor-*is*

e. If the name is feminine to it is added termination *-ae*. This termination is added irrespectively of the ending of the name.

Examples:

Name	E p i t h e t
Ivanova	Ivanov-*ae*
Olga	Olg-*ae*
Ross	Ross-*ae*
Compton	Compton-*ae*
Helena	Helen-*ae*

13. These epithets, substitutes of adjectives, *are not declinable* and therefore must not agree grammatically with generic names.

Examples:

Epithet—Personal name in Gentive Singular.	Epithet—adjectival form of a personal name.
Alchimilla Bungei	Dryadanthe Bungeana
Rubus Komarovii	Allium Komarovianum
Rubus Ruprechtii	Lonicera Ruprechtiana
Quercus Comptonae	

14. Epithets of species and of infraspecific taxa, which are compound adjectives, are formed by the combination into one word of:

    a. Two nouns joined together with the connecting vowel *i.*

       Examples:

         Hasta+folium=hast*i*folius, -a, -um

         Ramus+flos=ram*i*florus, -a, -um

         Latus+flos=later*i*florus, -a, -um

    b. Noun and adjective joined together with the connecting vowel *i.*

       Examples:

         Alternum+folium=altern*i*folius, -a, -um

         Ruber+flos=rubr*i*florus, -a, -um

         Longum+folium=long*i*folius, -a, -um

For further explanations concerned with formation of specific epithets which are compound adjectives cf. p. 127 et seq. of this guide.

§ . The name of an infraspecific taxon.

1. The formation of names of all infraspecific taxa follows the same rules as that of specific names.

2. The names of infraspecific taxa consist usually of one word. More rarely they consist of two words which are hyphenated or not.

       Examples:

         Galium verum var. *flore luteo.*

         Polygonum Thunbergii var. *hastato-trilobum.*

3. For denoting the rank of an ifraspecific taxon: variety (varietas), subspecies (subspecies) or form (forma, are used three methods. Of these only the first is in a current use, and the second and third were used *only* by early authors.

4. The first method is to interpolate the words variety, subspecies and form (or their abbreviations: *var., subsp.* [*ssp.*] and *f.* respectively) between the specific and infraspecific epithets.

Examples:

    Euonymus alata *var.* alata.

    Ranunculus cymbalaria *subsp.* sarmentosus.

    Leonurus sibiricus *f.* pallida.

5. The second method is applicable *only* to varieties. It was used by the early botanical writers who designated these infraspecific taxa by the letters of Greek alphabet.

    Examples:

    Euonymus alata *a* typica

    Euonymus alata *β* aptera

6. The third method is to place the varietal epithet immediately after specific one, the whole name being a trinomial.

    Examples:

    Euonymus alata typica

    Euonymus alata aptera.

2. The key.

§ 1. The nature and types of keys.

1. The key is a table in which the contrasting characters of different taxa of the same rank are arranged so as to facilitate their identification. For this purpose the distinguishing statements divide taxa into groups of first order, then of the second order, etc., until individual taxa are reached. Keys are one of the types of taxonomic literature and sometimes are written in Latin.

2. The keys are often called *analytical,* because they help to analyse the differences among the taxa belonging to one systematic group.

3. The key may be *natural* when it indicates natural relationships of the keyed taxa, or it may be *artificial* in which natural affinities are neglected, and are rather emphasized contrasts separating the taxa.

4. The key may be based on a *combination of all characters* of the plant (or the plant group), or only on *selected ones,* e.g. flowers, fruits or vegetative organs.

5. A *dichotomous* key is the one in which the contrasting statements are grouped in pairs, and give the choice only between two alternative characters. In *polychotomous* key the choice should be made among several characters. In current botanical work dichotomous keys are used preferably.

6. A pair of contradictory propositions in an dichotomous key is called a *couplet*, and each statement of a couplet is known as a *lead*.

7. The leads in the key are numbered or lettered in consecutive order, so that each couplet has its own number or letter. The numbering is better of the two methods, because in long keys there are not enough letters in the alphabet for lettering the leads.

8. With regard to their mechanical structure there are two types of keys, the bracket key and the indented key. Both types are used equally frequently.

9. An indented key is constructed in such a way that each successive subordinate couplet is indented under the one preceding it. Besides, the leads of one couplet are separated in the indented key, in some cases.

Below are given examples of indented key for five hypothetical taxa.

The form of the key is such as is adopted in the U. S. A.

The leads of the key are translated in the brackets.

Example 1: Key on the genus to section level.

A. Folia opposita [Leaves opposite].
   B. Inflorescentia densa, spiciformis [Inflorescence dense, spike-like].
      C. Flores lutei [Flowers yellow]   ...   *Taxon A.*
      C. Flores albi [Flowers white]   ...   *Taxon B.*
   B. Inflorescentia laxa, paniculata [Inflorescence loose, paniculate].   ...   ...   *Taxon C.*
A. Folia alterna [Leaves alternate].

      D. Flores lutei [Flowers yellow]   ...   *Taxon D.*
      D. Flores albi [Flowers white]   ...   *Taxon E.*

Example 2: Key on the specific level.

  A. Folia opposita [Leaves opposite].
    B. Folia utrinque glabra [Leaves on both sides glabrous].
      C. Folia integra [Leaves entire]   ...   *Taxon A.*
      C. Folia dentata [Leaves dentate]   ...   *Taxon B.*
    B. Folia subtus tomentosa [Leaves underneath tomentose]   ...   ...   *Taxon C.*
  A. Folia alterna [Leaves alternate]
    D. Folia utrinque glabra [Leaves glabrous on both sides]   ...   ...   *Taxon D.*
    D. Folia pilosa [Leaves downy]   ...   *Taxon E.*

10. In Europe are more commonly used bracket keys of the type represented below. In bracket key the successive couplets are not indented and continue on one line, and the leads of one couplet are never separated.

    Example 1: Bracket key for the same five hypothetical taxa on the genus to section level.

1 { Folia opposita	...	...	...	2
Folia alterna	...	...	...	4
2 { Inflorescentia densa, spiciformis		...		3
Inflorescentia laxa, paniculata		...		*Taxon C.*
3 { Flores lutei	...	...	...	*Taxon A.*
Flores albi	...	...	...	*Taxon B.*
4 { Flores lutei	...	...	...	*Taxon D.*
Flores albi	...	...	...	*Taxon E.*

Another variant of the same key.

    1. Folia opposita   2
    — Folia alterna   4
    2. Inflorescentia densa, spiciformis   ...   3
    — Inflorescentia laxa, paniculata   ...   *Taxon C.*

3. Flores lutei	...	...	...	*Taxon A.*
— Flores albi	...	...	...	*Taxon B.*
4. Flores lutei	...	...	...	*Taxon D.*
— Flores albi	...	...	...	*Taxon E.*

Example 2: Key on the specific level.

1 {Folia opposita	...	...	...	2
Folia alterna	...	...	...	4
2 {Folia utrinque glabra		...	...	3
Folia subtus tomentosa		...	...	*Taxon C.*
3 {Folia integra	...	...	...	*Taxon A.*
Folia dentata	...	...	...	*Taxon B.*
4 {Folia utrinque glabra		•••	...	*Taxon D.*
Folia pilosa	...	...	...	*Taxon E.*

Another variant of the same key.

1. Folia opposita	...	...	...	2
— Folia alterna	...	...	...	4
2. Folia utrinque glabra		...	...	3
— Folia subtus tomentosa		...	...	*Taxon C.*
3. Folia integra	...	...	...	*Taxon A.*
— Folia dentata	...	...	...	*Taxon B.*
4. Folia utrinque glabra		...	...	*Taxon D.*
— Folia pilosa	...	...	...	*Taxon E.*

11. The choice between the two types of keys is left to the discretion of the author of the botanical paper. In different countries, however, are traditionally preferred either bracket or indented keys. Generally speaking, bracket keys are more economic as of space, but indented keys are more easily followed.

12. The peculiarity of keys used in U. S. A. is that the numbers assigned to the couplets never repeat, even in indented keys. This is made in order to avoid confusion of leads. (Cf. example on pp. 16, 17

§ 2. Writing of a key.

1. The statements in the key are usually brief, technical, descriptions written in phrase form.

2. As the basis for further discussion here the key, as given on the p. 16 in example 1, is taken. This key is founded on three pairs of characters. According to the first pair of them (leaf arrangement) the plants are divided into two major groups, within which they are subdivided into minor groups on the basis of the type of inflorescence and color of the flowers, until the separate taxa are reached. The above mentioned characters are arranged in such a way as to draw a definite line between the two related groups.

3. When one starts to write the key, he should decide first which characters are common to all the taxa which are to be keyed, and which characters are contrasting. In a given example there may be two variants of this decision: (1) when one taxon, e.g. A, has a character not present in remaining four, and (2) when three taxa have a character not present in the other two.

4. In the first case the initial step will be to separate taxon A from the others in following way:

Indented key	Bracket key
A. Folia opposita *Taxon A.*	1. Folia opposita *Taxon* A.
A. Folia alterna	— Folia alterna ... 2

The second step will be to separate the remaining four taxa from each other in following way:

Indented key	Bracket key
A. Folia opposita *Taxon A.*	1. Folia opposita *Taxon A.*
A. Folia alterna	— Folia alterna ... 2
B. Inflorescentia densa, spiciformis.	2. Inflorescentia densa, spiciformis ... 3
C. Flores lutei *Taxon B.*	— Inflorescentia laxa, paniculata ... 4
C. Flores albi *Taxon C.*	3. Flores lutei *Taxon B.*
B. Inflorescentia laxa,	

paniculata.			— Flores albi	Taxon C.
D. Flores lutei	Taxon D.		4. Flores lutei	Taxon D.
D. Flores albi	Taxon E.		— Flores albi	Taxon E.

5. In the second case the first step will be to divide the given taxa into two groups as follows:

        Indented key   Bracket key

A. Folia opposita   1. Folia opposita ...   2
A. Folia alterna   — Folia alterna ...   4

and then the second step will be to separate in each of the groups one taxon from another in following way:

        Indented key   Bracket key

A. Folia opposita
B. Inflorescentia densa, spiciformis.
C. Flores lutei   Taxon A.
C. Flores albi   Taxon B.
B. Inflorescentia laxa, paniculata   Taxon C.
A. Folia alterna
D. Flores lutei   Taxon D.
D. Flores albi   Taxon E

1. Folia opposita ...   2
— Folia alterna ...   4
2. Inflorescentia densa, spiciformis ...   3
— Inflorescentia laxa, paniculata   Taxon C.
3. Flores lutei   Taxon A.
— Flores albi   Taxon B.
4. Flores lutei   Taxon D.
— Flores albi   Taxon E.

6. For the beginners in writing keys, it might be advisable to consider the following recommendations: (1) watch that the key is always strictly dichotomous; (2) watch that the statements in the two leads of each couplet are based on the contrasting characters; (3) to begin each lead of a couplet with the same word, and two consecutive couplets with different words; (4) not to employ in the key overlapping characters; the key based on such characters is undesirable since it complicates the identification and even may be misleading.

3. The diagnosis.
§ 1. The nature and types of diagnoses.

1. The diagnosis is a formal botanical description of any taxon made in order to publish this taxon effectively.
2. The plant(s) description in the diagnosis is primarily a qualitative one since this is a description of plant(s) organs with regard to their shape, size, color, texture, structure of their surface, etc.
3. The form of diagnosis varied considerably in different periods of history of descriptive botany. This may be learned by comparing early botanical writings with the modern ones. The diagnoses in the former ones differ from that in the latter ones in the form of presentation, style and vocabulary.
4. The initial form of diagnoses is, as it seems, the polynomial or the plant name joined with a brief description. These names were necessarily only specific diagnoses, since in Linnean times the diagnoses of genera were not known.

    Example of a polynomial :

    Allium scapo nudo tereti farcto, foliis semicylindraceis, staminibus corolla longioribus. (Gmelin, Flora Sibirica).
5. In contemporary descriptive botany, we may distinguish following three main types of diagnoses :
    1. Diagnosis of a genus, subgenus or section.
    2. Diagnosis of a species.
    3. Diagnosis of an infraspecific taxon.
6. There are some differences among all these types in minor details, but in general they are constructed according to one and the same plan, and have the same following essential parts :
    1. Initial paragraph consisting of the name of the taxon, the term defining its rank, and taxon's authority.
    2. Body of the diagnosis which is primarily morphological description of the taxon.
    3. Final paragraphs containing the discussion of the affinities of new taxon, its geographical distribution, ecology, and indica-

tions of its type specimens.

7. The main differences among the above mentioned types of diagnoses are as follows :

  a. In all types of them, except the diagnosis of species, the term describing taxon's rank is repeated twice.

    Examples :

    Genus Criptocodon Fed. *genus* novum.

    Sectio Platyphyllae Fed. *sectio* nova.

    Glycine gracilis Skv. *species* nova.

    Aegopodium alpestre Ldb. *forma* macrocarpa *forma* nova.

  b. In the diagnoses of all taxa above the species level the habitats and geographical distribution are usually not indicated.

  c. In the diagnoses of genera are usually more detailly described flower and fruit characters.

§ 2.   Writing of a diagnosis.

1. General recommendations to the beginner in writing of diagnoses :

  A. It is especially important to indicate in the diagnosis, the rank of the taxon described and the purpose of description (publication of a new taxon, change of name, etc.); and, in case of new taxon it is very important to describe its position in the system of the group, and its affinities to and differences from its allies.

  B. The description of a taxon (primarily of a genus or species) my be more detailed or less detailed. This is up to the author. However, it is not advisable to make it too detailed. The best description should be succint yet comprehensive, and should enable the reader to : (1) make up a mental picture of described plant(s), and (2) find all characters for the immediate identification of new taxon. The discussion of relationships of the new taxon and its differences from its allies usually also helps to identify it. Likewise helps to identify new taxon placing of it in the key of the group.

2. The shortness of diagnosis depends on omitting of unnecessary details, as indicated in the preceding paragraph, and besides, on the conciseness of the style. This is based on the possibility of abbreviation of sentences of which the description consists. (Cf. pp. 62-64 and 141,142 of this guide).

The method of abbreviation of the whole description is illustrated below by the example of hypothetical specific diagnosis since this type of description is most frequently met with in the work of taxonomist.

Example : Two variants of the same diagnosis of a hypothetical species.

The first variant is written in full sentences. but words which can be omitted are placed in brackets.

[Planta] perennis [cum] rhizomate crasso. Caulis [est] 30 cm. altus [cum] ramis divaricatis [in] parte superiore longe villosus, inferne subglaber. Folia [sunt] alterna [cum] petiolis 2 cm. longis [et] laminis ovatis, ca. 3 cm. longis, 2 cm. latis, [in] apice acutis [in] basi rotundatis. Flores [sunt] axillares in apicibus ramorum, [cum] calyce 5 mm. longo [et] corolla 10 mm. longa. Sepala [sunt] viridia, inferne pilosa [in] apice glabra, acuta, 2 mm. longa. Petala [sunt] rosea [cum] ungue 5.5 mm. longo [et] limbo 4.5 mm. longo. Pistillum et stamina [sunt] exserta. Fructus [est] 20 mm. longus, capsularis, viridis [cum] pedicello 3 mm. longo.

The second variant is the same diagnosis in abridged form. The words that were previously placed in brackets, are omitted.

Perennis, rhizomate crasso. Caulis 30 cm. altus, ramis divaricatis, parte superiore longe villosus, inferne subglaber. Folia alterna, petiolis 2 cm. longis, laminis ovatis, ca. 3 cm. longis, 2 cm. latis, apice acutis, basi rotundatis. Flores axillares in apicibus ramorum, calyce 5 mm. longo, corolla 10 mm. longa. Sepala viridia, inferne pilosa, apice glabra, acuta, 2 mm. longa. Petala rosea, ungue 5.5 mm. longo, limbo 4.5 mm. longo. Pistillum et stamina exserta. Fructus 20 mm. longus, capsularis, viridis, pedicello 3 mm. longo. To the beginner, to whom it may be difficult to write at once in such concise style, it may be recommended to prepare diagnosis, at first, in full sentences and then to shorten it by crossing out certain words.

II. Description of the taxa on the genus—series level.

1. Diagram of the arrangement of the elements in the diagnosis of a taxon on the genus—series level :

    A. Taxon rank. Taxon name.   Authority (Taxon rank) nov.

    B. Body of the diagnosis (botanical description).

    C. Discussion of affinities.

    D. Type of the taxon.

2. Construction of the elements of the diagnosis.

    Element A. may be constructed according to the following examples :

    Genus ; Gen *Criptocodon* Fed. gen. nov.

    Section : Sect. *Platyphyllae* Fed. sect. nov:

    Series : Ser. *Petreae* Klok. ser. nov.

For information regarding nature and methods of formation of generic names cf. pp. 3-7 of this work.

The terms describing taxon's rank and that the taxon's name is newly published are usually abbreviated. This part of diagnosis has mostly following form :

    New genus=genus novum=gen. nov.

    New section=sectio nova=sect. nov.

    New Series=series nova=ser. nov.

The element C. Discussion of affinities is made usually only for genera and subgenera, but not for sections and series.

The element D. The type of new taxon must be selected and certainly indicated in the diagnosis.

Geographical distribution is usually not indicated in the diagnoses of taxa above the species. However, for the genera it is advisable to describe their distribution, especially if the genus is endemic or has an distinctive geographical range.

3. Examples of diagnoses :

a. Diagnosis of a hypothetical new genus.

Genus ___(name)___ (authority) gen. nov.

Inflorescentia simplex, bracteis magnis, floribus hermaphroditis. Sepala distincta. Petala parva, inconspicua. Stamina 5. Filamenta brevia. Pistillum 1, stylo apice 5-fido. Discus nectariferus deest. Fructus obovoideus, baccatus. Semina 3 (raro 1-2), megna, alveolato-reticulata.

Differt a genere ___(name in Nominative Singular)___ qui proximum disci nectariferi absentia.

Generis typus   (name)

b. Diagnosis of a hypothetical new section.

Sectio ___(name)___ (authority) sect. nov.

Calyx alato-costatus.   Tubus corollae calyce triplo longior.

Flores corymbosi.

Sectionis typus   (name)

c. Diagnosis of a hypothetical new series.

Series ___(name)___ (authority) ser. nov.

Inflorescentia densa, floribus sessilibus.   Corolla pubescens.

Plantae palustres.

Seriei typus   (name)

I₁₁. *Description of a species.*

1. Diagram of the arrangement of elements in the diagnosis of a species.

A. Generic and specific names.   Authority. sp. nov.

B. Body of the diagnosis (botanical description).

C. Habitat.

D. Geographical distribution.

E. Affinities.

F. Type of the species.

2. Explanations about the sources of specific epithets, their nature and formation are given on pp. 7-14 of this work.

3. The terms that describe that the species or natural hybrid are newly described are as follows:

New species=species nova=sp. nov., spec. nov. or nova sp.

New hybrid=hybrida nova=hybr. nov.

These terms are interpolated after the authority of the new taxon.

Examples:

Valeriana leiocarpa Kitagawa sp. nov.

Populus girinensis Skv. sp. nov.

Potamogeton nitens Web. hybr. nov.

4. Thus the above examples demonstrate how is constructed element A. in the diagnosis of a new species.

5. In accordance with botanical tradition, morphological description of a species is made in following sequence:

(1) General habit.

(2) Root (or other underground organ).

(3) Stem.

(4) Leaves.

(5) Inflorescence (if present).

(6) Flower.

(7) Fruit

(8) Seed.

6. In conformity with this scheme, below are given the detailed plans for morphological description of herbaceous and woody species. The examples illustrating these plans are arranged in such a way as to demonstrate peculiarities of description of different types of organs in different taxonomic groups of plants.

Further discussion of "how to write diagnosis of a species" must be prefaced by following notes:

It should be clearly understood that the plans given below are not stereotyped forms for plant description, but merely general and suggestive examples. In practice they are always variously modified and adapted to the peculiarities of each plant which is described. This is because in the diagnoses are *on the first place described the characters on which is based the classification of the given group.* All other characters are described in the second place, or, even, not described at all if they are insignificant for the identification and taxonomic classification of the plant. For instance, linear form of leaves, common almost to all grasses, is usually not described in their diagnoses. In this group leaf form is described rather if it is not linear.

When describing particular plant organs it is recommended at first to describe their general character and then the details of the structure. In some cases, if the plant has some major distinctive character, that one is usually described in the first place in the diagnosis, to emphasize its significance.

Examples: The first sentence from the descriptions of certain species. Translation in brackets.

Planta tota glaberrima [All plant entirely glabrous]

Planta omnis sericeo-caana [All plant greyish-white sericeous]

Planta tota pilis ramosis hirsuta [All plant covered with coarse and stiff branched hairs].

Tota planta albo-tomentosa [All plant white-tomentose].

Planta tota pilis rufis longe villosa [All plant shaggy with long reddish, soft, straight hairs].

In case the herbarium specimen on which is based the description is incomplete, the absence of certain organs must be described in diagnosis. The method of doing this is explained on pp. 140, 141.

Description of herbaceous plant.

Plan of morphological description.

A. General habit (outer appearance) of the plant (small or large, tender or stout, erect, prostrate, climbing, etc.; glabrous or hairy, having some special typical characters or not. In some cases instead of general habit is described form of growth of the plant as: caespitose or non-caespitose, etc., or plant's longevity as: annual, biennial, limited perennial, perennial, semishrub).

Examples of the beginnings of diagnoses:

Planta annua [Annual plant]: Biennis [Biennial]; Perennis [Perennial]; planta aquatica [Aquatic plant]; Planta caespitosa [plant growing in tufts (caespitose)]; Perennis, dense caespitosa [Densely caespitose perennial]. Herba scandens, ca. 1 m. longa [Climbing herb, ca. 1 m. long]; Suffruticulus laxiuscule pulvinatus, ramis laxe floliosis [Somewhat loosely cushion-shaped small half-shrub, with loosely foliated branches].

B. Underground organs: bulbs, tubers, rhizomes or roots; their shape, size (large or small, thick or thin, etc.), position (shallow or profound, erect, oblique or horizontal), color, character of coating; producing or non-producing stolons, suckers, etc.

Examples:

a. Description of underground organs in general.

Radix fibrosa [Root fibrous]; Radix tenuis [Root slender]; Radix erecta, crassa [Root erect, thick]; Radix longe cylindrica, crassa, stolones graciles emittens [Root long, cylindrical, thick, sending forth slender stolons].

Radix intense fusca, lignosa, simplex vel apice valde ramosa, polycephala [Root intensely sombre brown, woody, simple or on the apex strongly branched, several-headed (i. e. with several shoots from a single crown)].

Rhizoma breviter repens [Rhizome short creeping].

Rhizoma crassum, multiceps [Rhizome thick, many-headed].

Rhizoma breve, crassum erectum, nigrum, densissime paleatum [Rhizome short, thick, erect, black, most densely beset with chaffs].

Rhizoma tenue, longe horizontaliter repens [Rhizome slender, long horizontally creeping].

Tuber sphaericum, 15-21 mm. diametro [Tuber sphaeric, 15-21 mm. across].

Tubera obconica, fusca, ca. 2 cm. longa, 1 cm. lata [Tubers reversed-conic, sombre brown, ca. 2 cm. long, 1 cm. broad].

Bulbus cylindricus, tunicis brunneis, densissme reticulatis [Bulb cylindrical, with most densely reticulate brown tunics].

Bulbus ovatus, fibris tenuibus dense vestitus, tunicis griseis [Bulb ovate, densely clothed with thin fibers, and with grey tunics].

Bulbus solitarius ovatus; tunicae subcoriaceae, in vaginam brevem productae [Bulb solitary, ovate; tunics subcoriaceous extended into short sheath].

Bulbus parvus, solitarius, tunicis atro-fuscis tectus [Bulb small, solitary, covered with dark sombre brown tunics].

b. Root crown or neck (in perennials and subshrubs only!) is described separately; are described size of the neck (long or short, thick or slender) and the structure of its surface (i. e. naked or covered with scales, chaffs, fibers, remains of leaves, etc.) If root neck is transformed into caudex (rootstock) this is specially indicated and the characters of the rootstock are described.

Examples:

Radix apice petiolis putridis foliorum radicalium dense vestita [Root on the apex is densely clothed with rotten petioles of radical leaves].

Radix apice squamis scariosis, griseo-brunneis, multiseriatis, obtecta [Root on the apex is covered with many-ranked, scarious, grey-brown scales].

Collum radicis residuis foliorum fibrosis dense vestitum [Root neck is densely clothed with fibrous remains of leaves].

Collum radicis nudum, saepe pluricaulis [Root neck naked, often several-stemmed].

Caudicis rami breves petiolorum reliquis obtecti [Branches of the rootstock short, covered with remains of petioles].

Caudex valde incrassatus, lignosus, pluriceps, breviramosus, reliquiis petiolorum vetustiorum sat dense tectus [Rootstock strongly thickened, woody, several-headed (i. e. with several shoots from a single crown); with short branches, covered densely enough with remains of old petioles].

Caudex elongatus, lignosus, cylindricus, inferne ad nodos radicans, superne caules 1-5 terminales vel laterales emittens [Rootstock elongate, woody, cylindrical, in the lower part at the nodes rooting, in the upper part giving forth 1-5 terminal or lateral stems].

Caudex crassus, lignosus, pluriceps, atro-fuscus [Rootstock thick, woody, several-headed, dark sombre brown].

C. Stem. Here may be two alternatives: (1) either plant has stem (plant caulescent), or (2) it has no stem (plant acaulescent).

(1) In caulescent plant is described stem, its form of growth (solitary or many, simple or branched, erect or prostrate, creeping, ascendent, etc.), form (straight or flexuose, etc.), form in cross section (rounded, triquetrous, etc.), structure of surface (smooth, striate, alate, etc.), character of indumentum (hairy, scaly, glandulose, prickly, etc.) or glabrous, color (if unusual!), character of branching, length and (if necessary) diameter in units of metric system.

Important note:

The Latin term for the stem is *caulis*. This term is used to describe stems of *all herbaceous plants with exception of grasses, sedges and rushes (Juncus)*. In descriptions of these plants the term *culmus*—culm is used.

Examples:

Grasses. Culmus 3-11 cm. longus, rigidulus, teres laevis [Culm 3-11 cm. long, somewhat rigid, terete, smooth].

Culmi recti, teretes, striati, 20-50 cm. alti, laxe puberuli [Culms straight, terete, striate, 20-50 cm. tall, loosely minutely pubescent].

Sedges. Culmi tenues, trigoni, scaberrimi, 15-35 cm. longi, basi vaginis purpureo-rubris aphyllis vestiti [Culms slender, with three angles, very scabrous, 15-35 cm. long, at base clothed with purple-red leafless sheaths].

Rush. Culmi rigiduli numerosi, 3-15 cm. alti e basi ramulosi [Culms somewaht rigid, numerous, 3-15 cm. tall, from the base bearing branchlets].

Culmi teretes leviter compressi, 30-60 cm. alti, 2-3 foliati, basi vaginis latis, fuscescentibus, vestiti [Culms terete, slightly compressed, 30-60 cm. tall, with 2-3 leaves, at the base clothed with broad sheaths turning somber brown].

Dicots. Caules a basi ramosi prostrati sulcati, ca. 25 cm. longi, parcissime farinosi [Stems from the base branched, prostrate, sulcate, ca. 25 cm. long, very sparingly mealy].

Caulis dense foliosus, glaber vel subglaber, 23-50 cm. altus [Stem densely foliated, hairless (glabrous) or subglabrous, 23-50 cm. tall].

Caulis elatus robustus, usque ad 30 cm. altus, ramosus, totus sessiliglandulosus [Stem tall, robust, up to 30 cm. tall, branched, all (covered) with sessile glands].

Caulis flexuosus, divaricato-ramosus, ca. 40 cm. altus, spinulosus [Stem zigzag, with spreading branches, ca. 40 cm. tall, spinulose.]

Caulis simplex vel in parte superiore ramosus, teres, vix sulcatus, 1-2 m. altus, glaberrimus tantum in inflorescentia glanduloso-pilosus [Stem simple or in the upper part branched, terete, scarcely sulcate, 1-2 m. tall, entirely glabrous only in the inflorescence glandular-pilose.]

(2) If the plant is acaulescent its description has a special character : The absence of stem is described by the expression *planta acaulis* which is usually placed at the very beginning of diagnosis, since it characterizes general habit (outer appearance) of the plant. After this are described the underground parts. The description of aerial parts of the plant has its own peculiarities, since acaulescent plants usually have rosulate radical leaves and floral scapes. The peculiarities are illustrated

by the following examples :

Example 1 :

Planta acaulis. Radix crassa, lignosa, verticalis. Folia radi-calia, rosulata, obovato-spathulata, 2-3 cm. longa, 0,5-0,8 cm. lata, utrinque tomentosa, margine integra. Scapus rectus, ca. 10 cm. longus, tomentoso-lanatus. Racemus 2-3 cm. longus. Pedicelli lana ti. bracteati. Flores caerulei, etc.

[Plant acaulescent. Root thick, woody, vertical. Leaves radical, rosulate, obovate-spathulate, 2-3 cm. long, 0,5-0,8 cm. broad, on the both surfaces tomentose, entire at the margin. Scape straight, ca. 10 cm. long tomentose-lanate. Raceme 2-3 cm. long. Pedicels lanate, bracteate. Flowers deep blue, etc.]

Example 2 :

Planta acaulis. Radices subfibrosae, crassiusculae—tenues, pallide fuscae. Folia pauca—numerosa, longe petiolata. Pedun-culi sub anthesin foliis breviores. [Plant acaulescent. Roots almost fibrous, from moderate thick to thin, pale somber brown. Leaves from few to many, long petioled. Peduncles at the flower-ing time shorter than the leaves].

So the diagnosis of acaulescent plant is written in the same way as that of caulescent plant only beginning from the description of inflorescence and flowers.

D. Leaves (arrangement: number: as leaves many or few; type of insertion as: leaves sessile or petiolate decurrent, clasping, etc.; shape; length and width in units of metric system; texture; presence or absence of indumentum and its type on each surface of the blade separately; color, type of vena-tion: special characters, e.g. unusual color, shape, indumen-tum, etc.) If the leaf is described part by part, then petiole, blade, stipules or sheaths, rachis and leaflets are described separately with regard to their size and structure.

Examples :

Group A. Leaves uniform in shape on whole of the plant.

(1)   Leaves are described as a whole.

Examples :

Fern.   Frons oblonga vel oblongo-elliptica, rigidula,   glaberrima, segmentis primariis lanceolato-triangularibus, acutis, secun-
dariis ovatis, crenatis [Frond oblong or oblong-elliptic, somewhat
rigid, entirely glabrous, with lanceolate-triangulate acute segments
of the first rank (and) ovate, scalloped segments of the second
rank].

Grass.   Folia late linearia, usque 1,4 cm. lata, longe acuminata,
plana, glabra vel sparse pilosa, marginibus scabris, revolutis.   Ligula
nulla. Vaginae foliorum   internodiis   breviores, glabrae, laeves
[Leaves broad linear, up to 1,4 cm. broad, long acuminate, flat,
glabrous or sparsely pilose, with scabrous, revolute   margins.
Ligule none. Sheaths shorter than internodes, glabrous, smooth].

Sedge.   Folia rigidula, longe attenuata, plana, 1-1,5 mm. lata,
culmo   breviora,   integerrima   [Leaves somewhat rigid,   long
attenuate, flat, 1-1,5 mm. broad, shorter than the culm, completely
entire].

Folia 1,5-2 mm. lata setiformi-convoluta,   longe attenuata,
flexuosa, culmo subaequilonga, superne scabra, margine obsolete
denticulata [Leaves 1,5-2 mm. broad, bristle-like longitudinally
rolled up, long attenuate, flexuose, nearly of the same length as
the culm, on the upper surface scabrous, at margin indistinctly
minutely toothed].

Dicots.   Folia sessilia, alterna, carnosa, obtusa, subdecurrentia,
glabra [Leaves sessile, alternate, fleshy, blunt (obtuse), subdecurrent,
glabrous].

Folia parva petiolata, ovata vel elliptica, obtusa, 0.25-0.75 cm.
longa, integerrima, supra viridia subtus dense squamis rufis obtecta
[Leaves small, petioled, ovate or elliptic, obtuse, 0,25-0,75 cm. long,
completely entire, on the upper surface green, underneath densely

34

covered with reddish scales].

Folia bipinnata laciniis margine inciso-dentatis, subtus arachnoideo-pilosa [Leaves duplicately pinnate partitions with incised-dentate at the margin, underneath cobwebby-pilose].

Folia oblonga pinnatim incisa vel pinnatifida, utrinque crispato-puberula, lobis integris [Leaves oblong pinnately incised or pinnately cleft, on the both surfaces crisped-downy, with completely entire lobes].

Folia multa, supra viridia, subtus albo-tomentosa, bipinnata, segmentis cuneiformibus, longe petiolulatis, plus minus profunde duplicato-dentatis [Leaves numerous, on the upper surface green, underneath white-tomentose, duplicately pinnate, with wedge-shaped long petiolulate segments, which are more or less profoundly duplicately dentate].

Folia ambitu orbiculari-cordata, palmatisecta, lobis vulgo biternatim fissis, dentibus terminalibus obtusis [Leaves in the outline orbiculate-cordate, palmately divided, with lobes which are usually biternately split, with obtuse terminal teeth].

Petiolus tenuis, ruber, folia 4-10 cm. longa, 2-9 cm. lata, initio parce hirsuta, mox utrinque glabra, ambitu ovata, basi cordata, profunde quinqueloba vel triloba, lobis serratis, longe acuminatis [Petiole slender, red, leaves 4-10 cm. long, 2-9 cm. broad, in the beginning sparingly hirsute, later on both surfaces glabrous, in the outline ovate, cordate at the base, profoundly five-lobed or three-lobed, with long acuminate, serrate lobes].

(2) Leaves are described part by part.

Note : this way describing leaves is used mostly when the leaves are pinnate or palmate, simple or compound.

Examples :

Fern. Stipites 5-8 cm. longi, nudi. Frondes 12-25 cm. longi, lanceolato-lineares, rigidulae, supra virides nitidae, subtus glaucae, pinnatifidae. Punnulae lineares 6-8 mm. longae, nervis 3-6 ramosis, margine crenulatae [Stipes 5-8 cm. long, naked. Fronds 12-25 cm. long, lanceolate-linear, somewhat rigid, on the upper surface green, shining, underneath glaucous, pinnately divided. Pinnules linear, 6-8 mm. long, with 3-6 branched nerves, at margin minutely scalloped].

Grass. Vaginae foliorum internodiis breviores, glabrae, laeves ; folia late linearia, 6-12 mm. lata, longe acuminata, plana, subtus laevia, superne ad nervos scabrida, radicalia caulinis breviora; ligula ad 5 mm. longa, elongata vel truncata [Leaf sheaths shorter than internodes, glabrous, smooth; leaves broad linear, 6-12 mm. broad, long acuminate, flat, underneath smooth, on the upper surface near the veins somewhat scabrous, radical shorter than the cauline ones; ligule up to 5 mm. long, elongate or truncate].

Dicot. Stipulae persistentes oblongo-ovatae, 9-14 mm. longae, 4-5 mm. latae. Petiolus gracilis, lamina aequilongus, glaber. Lamina ovata 50-70 mm. longa, 20-30 mm. lata, margine serrata, supra glabra, subtus secus venas adpresse pilosa [Stipules persistent oblong-ovate, 9-14 mm. long, 4-5 mm. broad. Petiole slender, equal in length to the blade, glabrous. Blade ovate, 50-70 mm. long, 20-30 mm. broad, serrate at the margin, on the upper surface glabrous, underneath appressed pilose along the veins].

Pinnate leaf. Folia 4-5 juga, imparipinnata, 5-9 cm. longa, petiolis dense pilosis, foliolis ellipticis vel obovatis, apice obtusis emarginatisve, 5-10 mm. longis, 3-7 mm, latis, adpresse sericeis [Leaves 4-5 paired, odd-pinnate, 5-9 cm. long, with densely pilose petioles, with elliptic or obovate on the apex obtuse and shallowly notched leaflets which are 5-10 mm. long, 3-7 mm. broad, appressed sericeous].

Palmate leaf. Petiolus longus, longe pilosus. Lamina foliorum orbiculari-reniformis, utrinque dense pubescens, trisecta, segmento medio obovato 3-dentato-lobato, lateralibus latioribus oblique rhomboideis, totis margine aculeatis [Petiole long, long pilose. Blade of the leaves orbiculate-kidney-shaped, on the both surfaces densely pubescent, divided into three segments with the middle segment obovate three-dentate-lobate, and with lateral segments more wide, obliquely rhombic, all at the margin prickly].

Group B. Leaves are not uniform in different parts of the plant, and are described type by type.

(1) Different leaves on different kinds of shoots.

Folia turionum setacea 3-7 cm. longa, marginibus asperis. Folia caulina opposita, linearia, 1.5-2.5 cm. longa et 0.8-1.5 mm. lata, aspera, basi vaginata [Leaves of the turions bristle-like, 3-7 cm. long, with rough margins. Cauline leaves opposite, linear, 1,5-2,5 cm. long and 0,8-1,5 mm. broad, rough, at the base sheathed].

(2) Different leaves in different parts of stem.

Examples :

Radical leaves. Folia radicalia sub anthesin jam decidua (or jam emarcida) [Radical leaves at the flowering time are already falling off (or are already withered)].

Folia radicalia sessilia, linearia, glabra, 6-9 cm. longa, 10-13 mm. lata, margine longe spinosa [Radical leaves sessile, linear, glabrous, 6-9 cm. long, 10-13 mm. broad, at the margin long spiny].

Folia radicalia longe petiolata, rotundata, basi subcordata, grosse subduplicato crenato-serrata, leviter setulosa [Radical leaves long petioled, rotundate, nearly cordate at the base, coarsely nearly duplicately scalloped-serrate, slightly minutely setose].

Cauline leaves. Folia caulina infima longe petiolata, biternata, segmenta sessilia vel breviter petiolata, ovata, acuminata, pinnato-incisa, marginibus profunde crenatis. Folia superiora subsessilia vel breviter petiolata [Lowermost cauline leaves long petioled, twice ternate, the segments sessile or shortly petioled, ovate, acuminate, pinnately cut, with profoundly scalloped margins. Upper leaves nearly sessile or shortly petioled].

Folia caulina infima post anthesin emarcida, folia caulina mediana et superiora sessilia, amplexicaulia, mediana linearia, basi paulo dilatata, apice acuminata, ad 7 cm. longa, ca. 0,4-0,6 cm. lata, superiora anguste-ovata, basi subcordata, 1-4,7 cm. longe, 0,7-1,2 cm. lata [Lowermost cauline leaves withering after flowering, middle and upper cauline leaves sessile, stem-clasping, middle linear, at base a little dilated, acuminate at the apex, to 7 cm. long, ca. 0,4-0,6 cm. broad, upper ones narrow ovate, nearly cordate at the base, 1-4,7 cm. long, 0,7-1,2 cm. broad].

E. Arrangement of flowers on the plant. Flowers solitary or many. If flowers are forming inflorescence characters of the latter (Form, size, many-or few-flowered, glabrous, pubescent, glandulose, etc.), bracts and pedicels (their form, size, texture, indumentum, etc.) are described.

Examples:

Flores terminales, singuli [Flowers terminal, one by one].

Flores solitarii, axillares [Flowers solitary, axillary].

Flos unicus, terminalis [Flower one (single), terminal].

Inflorescentia angusto-cylindrica, conferta, verticillis approximatis composita, fere aphylla [Inflorescence narrow cylindrical, dense (compact), composed of whorls set close together, nearly leafless].

Inflorescentia 4-6-flora, umbellata. Flores ca. 4 cm. in diametro. Pedicelli dense tomentosi, 8-12 mm. longi [Inflorescence 4-6-flowered, umbellate. Flowers ca. 4 cm. across. Pedicels densely tomentose, 8-12 mm. long].

Panicula ampla, laxa, 20-27 cm. longa, 10-14 cm. lata, rachi scabrida, ramulis scabris, erecto-patentibus [Panicle large, loose, 20-27 cm. long, 10-14 cm. broad, with somewhat scabrous rachis and scabrous, spreading branchlets].

Spiculae 3-5, angustae, cylindricae, rectae, superior mascula, ad 4 cm. longa, ceterae femineae, 2-4. 5 cm- longae, laxae [Spikelets 3-5, narrow, cylindrical, straight, the upper one male, up to 4 cm.

long, other female, 2-4, 5 cm. long, loose].

Umbella globosa plus minus 30-flora. Spatha ovata, acuminata, persistens. Bracteolae nullae, Pedicelli tenues ca. 1,5 cm. longi [Umbel globose (ball-shaped), more or less (approximately) 30-flowered. Spathe ovate, acuminate, persistent. Bracteoles none. Pedicels slender, ca. 1,5 cm. long].

Calathium 1, ovatum, 2.5 cm. longum. Pedunculus lanatus. Involucri phylla multiseriata, exteriora ovata, interiora linearia, albo-tomentosa, scarioso-membranaceo-marginata, apice spinulosa, receptaculum paleaceum [Capitulum 1, ovate, 2,5 cm. long. Peduncle woolly. Involucral bracts many-ranked, outer ovate, inner linear, white tomentose with scariously-membranous margins, on the apex spinulose, receptacle chaffy].

F. Flower. In flower, peduncle (in solitary flower) or pedicels (their size in unit of metric system, from, indumentum, etc.), calyx (size and structure, color, indumentum, etc.), corolla (size, form, color, etc.), appendices in calyx and/or corolla (tufts of hairs, scales, etc.), stamens and style(s) (size, form, indumentum, appendices, etc.), are described.

Examples :

Grass. Spiculae 3-6-florae, fuscescentes vel paulo violaceae, 4-7 mm. longae. Glumae steriles 1,5-2 mm. longae. Palea inferior elongato elliptica, 5-7-nervis, margine et apice membrancea, 2,6-3 mm. longa [Spikelets 3-6-flowered, turning somber brown or a little violet colored, 4-7 mm. long. Sterile glumes 1,5-2 mm. long. Lower palea elongate elliptic, 5-7-veined, at the margin and at the apex membranous, 2, 6-3 mm. long].

Monocot. Flos stellatus, albus. Tepala elliptica, 3 mm. longi, obtusa. Stamina linearia. Germen brevistipitatum subtriangulare. Stylus filiformis [Flower starry, white. Tepals elliptic, 3 mm. long, blunt (obtuse). Stamens linear. Ovary short-stalked, nearly three-edged. Style thread-like].

Rosaceous plant. Hypanthium obconicum, densissime tomentosum. Sepala brevia, deltoidea, acuta, patentia. Petala obovata. Styli cum staminibus subaequilongi, basi tomentosi [Floral tube (Hypanthium) reversed-conic, very densely tomentose. Sepals short, triangulate (deltoid), acute, spreading, Petals obovate. Styles and stamens nearly of equal length, at the base tomentose].

Legume. Flores 12-13 mm. longi, purpurei. Calyx campanulatus, 5 mm. longus, glaber, dentibus subulatis, tubo subaequilongis. Alae carina duplo breviores, planae. Carina apice rotundata. Vexillum obovatum, basi rubescens. Ovarium dense pilosum [Flowers 12-13 mm. long, purple. Calyx bell-shaped, 5 mm. long, glabrous, with awl-shaped teeth which are nearly equal in length to the tube. Wings twice as short as the keel, flat. Keel on the apex rounded. Standard petal obovate, at the base turning red. Ovary densely pilose].

Umbelliferous plant. Flores polygami, interiores masculi, exteriores hermaphroditi. Dentes calycini desunt, petala alba, obcordata, cacumine inflexo, 1 mm. longa. Stylopodium breviter conicum, styli patuli [Flowers polygamous, inner male, outer bisexual. Calyx teeth are absent, petals white, reversed cordate, with extreme top bent inwards, 1 mm. long. Stylopod short conic, styles spreading].

Dicot. Calyx parvus, dense villosus, dentibus triangularibus, acutis, ciliatis. Corollae purpureae, tubo incluso, labio inferiore declinato, tripartito [Calyx small, densely shaggy; with triangulate, acute, ciliate teeth. Corollas purple, with the tube comprised within (the calyx), with three-parted lower lip which is bent downwards].

G. Fruit (or sori in ferns). Type and mode of arrangement on the plant ; form, size, color, indumentum or other characters of surface, type of dehiscence, internal structure, are described.

Examples ;

Fern. Sori ampli biseriatim dispositi, substraminei. Indusia reniformia, plana [Sori large arranged in two rows, almost straw-yellow (colored). Indusia kidney-shaped, flat].

Sedge. Utriculi parvi, elongati, anguste ellipsoidei, 5-nervii, plani, substipitati, rostro recto truncato [Utricles small, elongate, narrowly ellipsoid, 5-veined, flat, almost stalked, with straight, truncate beak].

Cruciferous plant. Siliquae usque ad 2 cm. longae, erectae, tenuiter fusiformes, pilis stellatis tectae, 3-4-spermae [siliquas up to 2 cm. long, erect, lightly spindle-shaped, covered with stellate hairs, 3-4-seeded].

Ranunculaceous plant. Folliculi 5.5. mm. longi, lanceolati, hispidi, breviter rostrati [Follicles 5,5 mm. long, lanceolate, hispid, shortly beaked].

Prunoid plant. Drupa ca. 12 mm. longa, 11-12 mm. diam., paulo conica. Putamen conicum, 9-10 mm. longum, 5-6 mm. diam., sulcatum [Drupe ca. 12 mm. long, 11-12 mm. across, somewhat conic. Stone conical, 9-10 mm. long, 5-6 mm. across, furrowed].

Legume. Legumen albo-villosum, 4-7-articulatum, articulis suborbicularibus, 5-7 mm, longis et 4 mm. latis [Legume white-shaggy, 4-7-articulate, with almost orbicular segments which are 5-7 mm. long and 4 mm. broad].

Umbelliferous plant. Fructus maturus brunneo-violascens, oblongus, 0,3 cm. longus. Stylopodia nigra. Juga ferruginea. Vittae valleculares 3 in vallecula, comissurales 3-4 [Fruit at maturity turning-brown violet, oblong, 0,3 cm. long. Stylopods black. Ridges rusty. Vittae are 3 in a furrow, 3-4 on the surface of union].

Composit. Achenium 0,75-1 mm. longum, cinnamomeum, ovatum, striatum, dense glandulosum. Rostrum subnullum. Pappus pilis scabris, sordide albus [Achene 0,75-1 mm. long, cinnamon-colored, ovate, striate, densely glandulose. Beak almost none. Pappus (consists) of scabrous (rough) bristles, dirty white].

H. Seeds. Form, size, color, characters of the coat (smooth, pitted, rugose, etc., dull or shining): carunculae or other appendices and their characters, are described.

Examples :

Semina elliptica, parva, ad 1 mm. longa, pallide brunnea, nitida [Seeds elliptic, small, up to 1 mm. long, palely brown, shining]. Semina glabra, olivacea, minutissime rugulosa [Seeds glabrous, olive-green, very minutely somewhat wrinkled].

Semina brunnea, oblonga, ca. 2,5 mm. longa, ca. 0,5 mm. lata [Seeds brown, oblong, ca. 2,5 mm. long, ca. 0,5 mm. broad].

Semina parva, nigricantia, sub lente minute acutissime verrucosa [Seeds small, blackish, under a lens minutely very sharply warty].

Semina nigra, 2 mm. longa, minutissime punctulata [Seeds black, 2 mm. long, very minutely dotted].

6. Habitat. In the diagnosis this paragraph is designated by the term *habitatio*, abbreviated hab.

Examples of description of habitats :

In arenosis [On sandy places].

In paludibus [In marshy or in swampy places].

In saxosis [On rocky or stony places].

In ruderatis [On rubbish dumps].

In fruticetis [In shrub thickets].

Ad ripas [At riverside].

In declivibus siccis (humidis, argillosis, etc.) [On dry (moist, clayey, etc.) slopes].

In stepposis aridis (alkalinis, montanis, etc.) [In dry (alcaline, mountain, etc.) steppes].

In herbidis subalpinis (ripariis, silvaticis, etc.) [On subalpine (riverside, forest, etc.) grassy places].

In rupibus muscosis (apricis, umbrosis, etc.) [On the mossy (open sunny, shady, etc.) cliffs].

In silvis montanis (frondosis, acerosis, etc.) [In the mountain (deciduous, coniferous, etc.) forests].

In pratis siccis (humidis, ripariis, etc.) [On the dry (moist, riverside, etc.) meadows].

In aquis stagnalibus (currentibus, lente currentibus, etc.) [In standing (flowing, slowly flowing, etc.) waters].

If species occurs in different ecological conditions this is described in following way :

Examples :

In paludibus et aquis stagnalibus vel lente currentibus [In swamps and standing or slowly flowing waters].

In pratis humidis vel herbidis ad ripas fluviorum [In damp meadows or in grassy riverside places].

In silvis frondosis vel fruticetis [In deciduous forests or in shrub thickets].

In pratis siccis vel in stepposis [In dry meadows or in steppes].

7. Geographical distribution. This paragraph in diagnosis is termed in Latin as *distributio,* abbreviated *distrib.,* or *area geographica* [geographical range], abbreviated, *area geogr.* or *ar. geogr.* Early writers used to write *patria* [native country]. If the species is endemic, in geographical distribution this is described by the term *endemica.*

Examples :

Asia Media [Central Asia]; America borealis [northern America]; Oriens Extremus [Far East]; Regio Ussuriensis [Ussuri Region]; Regio Amurensis [Amur Region]; Siberia occidentalis [western Siberia]; China australis [southern China]; Insula Dagelet (endemica) [Dagelet Island (endemic)].

8. Affinities. In this paragraph are discussed the differences of the new species from and its affinities to its allies. This discussion is usually presented in the following form:

(1) Differt a specie (or speciebus, if many)......[1]......[2]......
[From the species (or species, if many)......[1]......is different in......[2]......].

(2) Proxima] ...[Affinis] ...$^{(1)}$..., sed...$^{(2)}$...[distincta or differt] [Is nearest to ...$^{(1)}$..., but differs (from it) in...$^{(2)}$...].

(3) Valde affinis...$^{(1)}$...[ex qua (quo, quibus) or a qua (quo, quibus)] differt....$^{(2)}$...[Is strongly allied to ...$^{(1)}$...from which differs in...$^{(2)}$...].

The above models are the formulae for description of affinities in which should be inserted only the necessary data which are indicated by the numbers in parenthesis. These numbers mean :

Model (1), (1) name(s) of related species in Nominative case.

Model (2), (1) name(s) of related species in Dative case.

Model (3), (1) name(s) of related species in Dative case.

In all models (2) means enumeration of distinctive characters in Ablative case.

It should be noted that only one of these models must be used at a time. Example of the usage of the formula No. 1:

Differt a speciebus *G. effusa* Kit. et *G. leptolepis* Ohwi culmis gracilibus, paniculis plus divaricatis et glumis sterilibus pilosis [Differs from the species *G. effusa* Kit. and *G. leptolepis* Ohwi in more slender culms, in more divaricate panicles and in pilose sterile glumes].

Note : The pronoun *qua* is used when the plant name preceding to this pronoun is feminine. In case if the plant name is masculine or neuter the pronoun *quo* is used and if the names are several is used the plural form of the pronoun—*quibus*.

Examples :

Valde affinis A. silvaticae [a qua / ex qua] differt foliis pilosis.

Valde affinis A. silvatico [a quo / ex quo] differt foliis fistulosis.

Valde affinis A. silvatico et A. pratensi [a quibus / ex quibus] differt foliis fistulosis.

9. Type of the species. The type of new species is named and described in the special paragraph at the end of diagnosis. Below is given a model formula for writing this paragraph :

1. Variant.

Specimina examinata: ......[1]...[2]...prope...[3]..., legit...[4]...... [5]..., typus. In herbario...[6]... conservatur (or simply typus in herbario......). [Examined specimens: ...[1]..., ...[2]...near...[3]..., collecte...[4]...[5]..., type. Preserved in the herbarium of...[6]...(or simply type in the herbarium of......)]

2. Variant.

Typus speciei :...[1]...[2]...prope...[3]...[5]...[4]...legit, in herbario... [6]...conservatur [Type of the species:...[1]...[2]...near...[3]...[5]...[4]... collected, preserved in the herbarium of...[6]....]

The necessary data which are indicated by the numbers in parenthesis and which must be inserted in the above forms are as follows :

(1) country and province where the specimen was collected.

(2) habitat of the specimen.

(3) Name of the town, village, lake, mountain, etc. near which the specimen was collected.

(4) Collector's name.

(5) Date of collecting specimen.

(6) Name of the herbarium where the specimen is preserved.

The way in which the above formulae are used in botanical writings is illustrated by the following examples.

Examples:

Specimina examinata: Manshuria, prov. Heilungkiang, in pratis humidis prope Harbin, legit W. Wang, Sept. 12, 1937, typus. In herbario Academiae Scientiarum Sinicae conservatur [*Examined specimens*: Manchuria, Heilungkiang province, in the wet meadows near Harbin, collected W. Wang, 12-th Sept., 1937, type. Preserved in the herbarium of the Academy of Sciences of China.]

Specim. exam. Manshuria:

Prov. Mu-tan-chiang: Er-tou-he-tzu (M. Takenouchi et Y. Watanabe Sept. 6. 1938 — Typus) *Examined specimens*, Manchuria: Mu-tan-chiang province: Er-tou-he-tzu (M. Takenouchi and Y. Watanabe 6-th Sept., 1938 — type)].

certain authors instead of specimina examinata write specimina visa
specimens seen. In this case often the citation of type and of
the examined specimens are given in separate paragraphs.
Examples:
Specimina examinata. Transcaucasia: Mingrelia, Djwari in pratis
alpinis 25 VIII 1893 N.Alboff; Mingrelia m. Kwira pratis alpinis in
declivibus australibus 16 VIII 1890 No. 860 N. Kusnezov.
Typus: in Abchasia ad rupes calcareas, in angustiis fluminis
Okum, alt. circa 300 m No. 2 19 IV 1893 (in Herb. Ac. Sc. URSS).
[*Examined specimens.* Transcaucasus: Mingrelia, Djwari in alpine
meadows 25 Aug. 1893 N. Alboff; Mingrelia mountain Kwira in
alpine meadows on southern slopes 16 Aug. 1890 No. 860 N. Kusnezov.
Type: in Abkhasia on the limestone cliffs, in the gorge of the
river Okum, at elevation about 300 m No. 249 April 1893 (in Herbarium of Acad. Sci. of URSS)].
Typus: in trajecto Scharodil inter Achsu et Schemacha, in locis
graminosis 2 V 1908 fl. lg. G. Woronow. In Herb. Inst. botan. nom.
ac. V.L. Komarovii Ac. Sc. URSS in Leningrad conservatur.
Specimina visa: Tbilisi, Saguramo, Kodzhory, Schiraki, Ivanovka,
inter Geoglar et Ketany [Type: in the mountain pass Sharodil
between Akhsu and Shemakha, on grassy places 2 May 1908 flowering
collected G. Woronow. Preserved in the Herbarium of the Komarov's Botanical Institute of the Acad. Sci. of URSS in Leningrad].
Specimens seen: Tbilisi, Saguramo mountain range, Kodzhory,
Shiraki, Ivanovka, between Geoglar and Ketany].
Typus: prope Tbilisi in monte St. Davidi 19 V 1961 fl. leg.
Owerin. In Herb. Inst. bot. nom. ac. V.L. Komarovii Ac. Sc. URSS
in Leningrad conservatur.
Specimina visa numerosa ex omni speciei area geographica.
[*Type*: near Tbilisi on the St. David's mountain 19 May 1961
flowering collected Owerin. Preserved in the Herbarium of the
academician V. L. Komarov's Botanical Institute of the Acad. Sci. of the

URSS in Leningrad.

Specimens seen are numerous from the whole geographical area of the species]. On the other hand some authors join together in one paragraph the citation of type specimen and the description of the habitat of a new species.

Example:

Habitat in pratis humidis prope pag. Ivanovka ubi 5 VI 1950 a Borisov lecta. Specimina authentica in Herb. Inst. Bot. Acad. Sc. URSS in Leningrad conservatur [It grows in wet meadows near the village Ivanovka where it was collected 5 June 1950 by Borisov. Authentic specimens are preserved in the Herbarium of the Botanical Institute of Acad. Sci. URSS in Leningrad].

Below I continue to give the examples of citation of type specimens. These examples show how this citation is variously modified by different authors.

Examples:

Typus: prope Ja-Tschshou 7. IV. 1893 G. N. Potanin, in Herb. Inst. Bot. Ac. Sc. URSS conservatur [*Type*: near Ya- chou 7 April, 1893 (collected) G. N. Potanin, preserved in the Herbarium of the Botanical Institute Acad. Sci. URSS].

Typus: Transcaucasia Terter, Isty-su, in pratis subalpinis, 28. VII. 1929, fl. L. Utkin, in Herb. Universitatis Mosquensis conservatur [*Type*: Transcaucasus Terter, Isty-su, in subalpine meadows, 28 July, 1929, flowering (collected) L. Utkin, preserved in the Herbarium of Moscow University].

Typus: Asia media. Montes Karatau, in Herb. Univers. As. Med. conservatur [*Type*: Central Asia. Karatau mountains, preserved in the Herbarium of the University of Central Asia].

Typus: In vicinitate pag. Sonskoje, prata stepposa 25. V. 1910, fl., Smirnov (Herb. Inst. Bot. Ac. Sc. URSS) [*Type*: in vicinity of the village Sonskoie, on steppe meadows 25 May, 1910, flowering, (collected) Smirnov (Herbarium of the Botanical Institute Acad. Sci. URSS)].

Prov. Feng-t'ien: in monte He-shang-shan prope Chin-chou (M. Kitagawa Jul. 24, 1930—Typus) [*Feng-t'ien province*: on the He-shang-shan mountain near Chin-chou (collected) M. Kitagawa 24 July, 1930—Type)].

Prov. Jehol: In silvis montis Wu-ling-shan 500-900 m (N.H.K., 1 Sept. 1933, typus in herbario Universitatis Imperialis Tokyoensis) [*Jehol province*: in the forests of Wu-ling-shan mountain (altitude) 500-900 m (collected) (Nakai, Honda, Kitagawa 1 Sept. 1933, type in the Herbarium of Tokyo Imperial University)].

Typus: Regio Primorsky, distr. sinus Olga, ad fl. Avvakumovka, in rupibus, 11. VIII 1951, leg. D.P. Vorobjov (in Herb. Hort. bot. princ. Ac. sci. URSS, Mosqua) [*Type*: Primorskii region, Bay of Olga district, at the river Avvakumovka on the cliffs 11 August 1951, collected D. P. Vorob'ev (in Herbarium of the Principal Botanical Garden of Acad. Sci. USSR, Moscow).

Speciei typus: Asia orientalis, prov. Austro-Ussuriensis, ad ripas fl. Sidemi in districtu Possiet 21. IX. 1933, leg. V. L. Komarov [*Type of the species*: Eastern Asia, Southern Ussurian province, on the banks of river Sidemi in Possiet district 21 September, 1933, collected V. L. Komarov].

Typus sectionis: Veronica ciliata Fisch. [*Type of the section*: Veronica ciliata Fisch.]

Typus seriei: Andrachne rotundifolia C.A.M. [*Type of the series*: Andrachne rotundifolia C.A.M.].

Generis typus: Spirostegia bucharica (B. Fedtsch.) Ivanina (*Type of the genus*: Spirostegia bucharica (B. Fedtsch.) Ivanina].

Since today few people put every part of the description in Latin, I give below, for the guidance of the readers, several examples of the citation of type specimens in English.

Examples:

Type in the U. S. National Herbarium No. 2,306,491, collected in Leipzig Ave. Pond, Germania, Atlantic Co. New Jersey, June 7, 1959, by Frank Hirst.

Type in the Gray herbarium collected on a slope in open pinyon forest, 1-2 miles southwest of Pablillo, Nuevo Leon, Mexico, July 21, 1958, D.S. Correll and I.M. Johnston 1941. Isotype in the Lundell Herbarium, Texas Research Foundation.

Type: Craig Mts., Nez Perce Co., Idaho, Henderson 2770 (US). Isotypes [RM, MSC, WS (incl. photo of type)].

Type: Rocky Mts. Lat. 49 N. Alt. 6-7000 ft. above the sea. 1861. Lyall s.n. (GH).

Solomon Islands: Guadalcanal: Nalimbu River, F. S. Walker B. S. T. P. 4 (TYPE).

Northeast New Guinea: Aiyura, *L. S. Smith* N. G. F. 1019, Oct. 1944 (Type, leaves and fruit) mountain rain forest.

Vanua Levu: Mathuata: summit ridge of Mt. Numbuiloa, east of Lambasa, alt. 500-590 m., Nov. 6. 1947, *Smith 6520* (GH, US Type) (in dense crest forest).

CHINA: Yunnan: Li-kiang, *R.C. Ching 20646* (Type, A).

CHINA: Yunnan: Wei-se, *C. W. Wang 64164* (Type, A).

For all of these citations it is typical that there is tendency to make them as short as possible. In order to do this are used abbreviations and conventional signs: the names of herbaria where the type specimen is deposited are abbreviated, the collector's name and the number of the specimen are italicised.

The data for the citation of examined specimens are taken from the labels. But, sometimes, the labels are incomplete. How to write citations of specimens in this case is explained on pp. 140,141 of this guide.

In addition to all that was said before about the writing of the diagnosis of species below are given several examples of somewhat standardized diagnoses of hypothetical herbaceous species belonging to three main groups of the plant kingdom.

Examples:

1. Fern.

Name. Authority sp. nov.

Rhizoma validum. Frondes ad......m. longi. Stipes lamina brevior (longior, etc.), dense paleatus (glaber, etc.). Lamina ambitu ellipticooblonga (lanceolata, etc.), duplicato-pinnata. Segmenta primaria lanceolato-linearia, acuminata, secundaria ovata, crenulata. Sori secus venas dispositi. Indusium reniforme, membranaceum, persistens. Sporae subglobosae, verrucosae.

Note: The words in parenthesis must give an idea of how this and other descriptions might be changed, if necessary.

[Name. Authority sp. nov.

Rhizome thick (or thin, etc.). Fronds up to......m. long. Stipe shorter (or longer, etc.) than blade, densely chaffy (or glabrous, etc). Blade in the outline elliptic-oblong (or lanceolate, etc.), duplicately pinnate. Segments of the first rank lanceolate-linear, acuminate, of the second rank ovate, crenulate. Sori are arranged along veins. Indusium kidney-shaped, membranous, persistent. Spores are subglobose, warty.]

Hab......

Area geogr......

Affinis........., sed differt...... .....

Specimina examinata...............

2. *Grass* (*Monocot.*)

Name. Authority sp. nov.

Perennis. Rhizoma longum, tenue (breve, validum, etc.). Culmus a basi ascendens, dein rectus usque ad...m. altus. Folia linearia, plana, tenera (convoluta, crebra, etc.),...mm. lata. Lamina supra scabridula, subtus laevissima, apicem versus acuta (obtusata, etc.) Ligula elongata,...mm. longa. Vaginae laeves (scabrae, etc.). Panicula ampla (contracta, etc.), ramulis capillaribus, patentibus, (erectis), scabris (laevibus). Spiculae longe (breviter) pedicellatae,...-florae,... mm. longae. Glumae steriles carinatae (ecarinatae), aristatae (non ariastatae). Paleae apice bifidae (integrae, acuminatae, aristatae, etc.),...mm. longae.

[Perennial. Rhizome long, slender (or short, thick, etc.) Culms at the base ascendent thereafter straight, up to...m. tall. Leaves linear, flat, tender (or convolute, thick, coarse, etc.),...mm. broad. Blade on the upper surface slightly rough, underneath entirely smooth, towards apex acute (or obtusate, etc.) Ligule elongate,...... mm. long. Sheaths smooth (or rough, etc.) Panicle broad [large] (or contracted, etc.), with hair-like, spreading (or erect), rough (or smooth) branches. Spikelets long (or short) pedicelled, with... flowers,...mm. long. Sterile glumes keeled (or not keeled), awned (or not awned). Paleas on the apex divided in two parts (or non-divided, acuminate, awned, etc.),...mm. long.]

Hab......

Area geogr..............

Affinis..................., sed differt.........................

Specimina examinata...

3. *Sedge* (*Monocot.*)

Name. Authority sp. nov.

Perennis. Planta virescens, caespitosa. Culmi laeves (scabri), subtrigoni...cm. alti. Folia plana,...mm. lata, subito (leniter) acuminata (acuta, obtusa, etc.), culmo aequilonga (breviora, longiora, etc.) Spiculae numero..., distantes (congestae). Spicula terminalis mascula (androgyna, etc.), lineari-lanceolata (obovata, clavata, etc.), ...cm. longa. Squamae angustae (latae), pallidae. Spiculae inferriores femineae, hemisphericae (globosae, ellipsoideae, etc.),...cm. longae. Squamae ovatae (lanceolatae, ellipticae, etc.) subferrugineae (brunneae, fuscae, stramineae, etc.). Utriculi lanceolati, obtusotrigoni....mm. longi, flavescenti-virides (brunneae, stramineae, etc.), apice rostro conico (cylindrico, etc.), bidentato (non dentato, etc.).

[Name. Authority sp. nov.

Perennial. Plant turning green, growing in tufts. Culms smooth (or rough), almost triquetrous,...cm. tall. Leaves flat,...... mm. broad, suddenly (or gradually) acuminate (or acute, obtuse, etc.), equal in length to the culm (or shorter or longer than the culm). Spikelets......in number, remote (or crowded together). Terminal spikelet male (or androgynous, etc.), linear-lanceolate (or obovate, clavate, etc.)...cm. long. Scales narrow (or broad), pale. Lower spikelets female hemispherical (or globose, ellipsoidal, etc.), ...cm. long. Scales ovate (or lanceolate, elliptic, etc.) nearly rusty (or brown, sombre brown, straw-yellow, etc.) Perigynia lanceolate, obtusely-triquetrous, ...mm. long, yellowish-green (or brown, straw-yellow, etc.), on the apex with conical (or cylindrical, etc.), bidentate (or non-dentate, etc.) beak].

Hab...............

Area geogr.................

Affinis..............., sed differt......................

Specimina examinata....................................

4. *Dicot.*

Name. Authority. sp. nov.

Perennis (annua, biennis). Planta habitu diffuso laxe patenteque ramosa (habitu compacta ramis erectis, etc.) Radix valida (debilis, tenuis, etc.), apice cicatricibus caulorum notata. Caulis erectus (prostratus, scandens, etc.), glaber (pilosus, hirsustus, glandulosus, etc.), ca...cm. altus. Folia elliptica (ovata, lanceolata, linearia, etc.), petiolata (sessilia),...cm. longa,...cm. lata, utrinque glabra (pilosa, villosa, etc.), margine subdenticulata (dentata, serrata, crenata, etc.) Inflorescentia terminalis (axillaris). Pedicelli ca.... mm. longi. Bracteae lanceolatae, cordatae, ellipticae, subulatae, etc.), ca...mm. longae. Calyx...mm. longus, dentibus triangulatoovatis (subulatis, lanceolatis, etc.), acutis (acuminatis, obtusatis, truncatis, etc.) Corolla ca...mm. diametro, lobis patentibus. Stamina filamentis complanatis (teretibus), basi dilatatis. Antherae ellipticae (ovatae, globosae, etc.), ...mm. longae. Ovarium ellipsoideum (globosum, ampulliforme, etc.), stylus filiformis, stigma capitatum. Fructus ovoideus (ellipsoideus, globosus, etc.), erectus (nutans),...cm. longus,...cm. crassus.

[Name. Authority. sp. nov.

Perennial (or annual, biennial, etc.). Plant of the loose outer appearance with loosely spreading branches (or compact in outer appearance with erect branches, etc.) Root thick (or weak, thin, etc.) on the apex with stem-scars. Stem erect (or prostrate, climbing, etc.), glabrous (or pilose, hirsute, glandulose, etc.), ca....cm. tall. Leaves elliptic (or ovate, lanceolate, linear, etc.), petiolate (or sessile),...cm. long,...cm. broad, on the both surfaces glabrous (or pilose, villous, etc.), along the margin subdenticulate (or dentate, serrate, scalloped, etc.) Inflorescence terminal (or axillary). Pedicels ca...mm. long. Bracts lanceolate (or cordate, elliptic, awl-shaped, etc.), ca...mm. long. Calyx...mm. long, with triangulate-ovate (or awl-shaped, lanceolate, etc.), acute (or acuminate, obtusate, truncate, etc.) lobes. Corolla ca...mm. across, with spreading lobes. Stamens with flattened out (or terete) filaments, dilated at base. Anthers elliptic (or ovate, globose, etc.),...mm. long. Ovary ellipsoidal (or globose, flask-shaped, etc.), style thread-like, stigma capitate. Fruit ovoid (or ellipsoidal, globose, etc.) erect (or nodding),...cm. long,... cm. thick].

Hab........................

Area geogr.............................

Affinis..............., sed differt.......................

Specimina examinata..........................

Descriptions of aquatic, parasitic and saprophytic herbaceous flowering plants have certain peculiarities which are illustrated by the following examples.

Examples :

Aquatic plants.

Herba aquatica, foliis natantibus, elliptico-oblongis, apice et basi obtusis, nervis parallelis utrinque 4 percursis venulis transversis plurimis, etc. [Aquatic herb with floating, elliptic-oblong, obtuse at base and apex leaves, with four parallel veins on both surfaces and with numerous transverse veinlets, etc.]

Caules immersi elongati hic inde parce ramosi dense foliati, folia, etc. [Stems immersed elongate locally sparingly ramose, densely foliated, leaves, etc.]

Herba parva submersa monoica, caulibus tenuibus basi limum permeantibus et radicantibus, superne effuse ramosis, ramis gracilibus, dense foliatis patente adscendentibus, etc. [Small submerged monoecious herb with slender stems which at base are passing through the mud and rooting, in the upper part are widely ramose, with slender, densely foliated, ascendent-spreading branches, etc.]

Planta annua, tenera, debilis, fragilis, obscuro-viridis, demersa. [Annual, tender, weak, fragile, dark green, submerged plant].

Parasitic plant.

Planta herbacea, glabra, parasitica, achlorophyllea, violascens, humilis, fere omnis humo immersa, dense pulvinatim multicaulis, caulibus teretibus, etc. [Herbaceous, glabrous, parasitic, achlorophyllaceous, turning violet, low-growing plant almost all imbedded in the soil densely cushion-like many-stemmed, with terete stems, etc.]

Saprophytic plant.

Herba saprophytica, perennis. Radices numerosae, breves, carnosae, crassae, arcte confertae, Caulis 8-15 cm. altus, carnosus, glaber, pallidus. Squamae (folia) sessiles, alternae, late ellipticae usque late ovatae, plurinerviae, margine integrae, 9-15 mm. longae, 6-10 mm. latae, florem versus capituliformi-coarctatae in petala transeuntes. Flos albus, etc. [Saprophytic perennial herb. Roots numerous, short, fleshy, thick, densely crowded. Stem 8-15 cm. tall, fleshy, glabrous, pale. Scales (leaves) sessile, alternate, broad elliptic up to broad ovate, many-veined, entire at the margin, 9-15 mm. long, 6-10 mm. broad, toward flower head-like close set changing into petals. Flower white, etc.]

Description of a woody plant.

1. Descriptions of the woody plants are made according to the same general plan as of the herbaceous plants.

2. The differences between the description of a woody and of a
herbaceous plant are in that the beginning of the diagnosis of a
woody plant is written in a different way. This is illustrated
by the examples given below.

Plan of a description of a woody plant.

A. General habit [outer appearance] of the plant (tree, shrub,
undershrub; small or large; form of head, or form of growth;
height and diameter of trunk), color and character of bark on
trunk, branches and branchlets.

Examples:

1. *Tree.*

Arbor alta; cortex albus charataceo-dirumpens. [Tall tree;
bark white paper-like-exfoliating].

Arbor 30 m. alta; cortex trunci griseo-fuscescens papyraceo-solu-
tus; lenticellae lineares. [Tree 30 m. tall; bark of the trunk grey-
turning somber brown paper-like-exfoliating; lenticels linear].

Arbor 20-metralis; coma ovoidea; cortex squamatus. [Tree
20 m. tall; head ovoid; bark scaly].

2. *Shrub.*

Frutex volubilis [Twining shrub]. Frutex scandens [Climbing
shrub]. Fruticulus humilis [Low small shrub].

Frutex glaber, ramis oppositis gracillimis, erecto-patentibus,
teretibus, cortice viridi, nigro-punctato. [Glabrous shrub with very
slender opposite, erect-spreading, terete branches, with green, black-
dotted bark].

B. Branches and suckers. In both of them are described characters
of stem, bark, indumentum, leaves, etc. The characters of
annotinous and hornotinous branches are described separately,
e.g. buds, their form, size, etc.

Examples:

Rami glabri, castanei, parce albo-lenticellati; gemmae minutae
crasse ovatae, perulis latis, apice nigris squarrosis. [Branches
glabrous, chestnut-brown, sparingly white-lenticellate; buds small

thick ovoid, with broad, on the apex black, squarrose scales].

Ramuli tenuiusculi, primum angulati, fulvo-tomentosi, demum fusco-grisei, nitidi, indistincte lenticellati; gemmae grandes, fusiformes, acuminatae. [Branchlets somewhat slender, at first angled, tawny-tomentose, later somber brown-grey, glossy, indistinctly lenticellate ; buds large, spindleshaped, acuminate].

Ramus juvenilis glandulosus, lenticellis albis punctulatus [Young branches glandulose, minutely punctate with white lenticels].

Caules annotini teretes, cinnamomei, dense hirsuti et aculeati; ramuli floriferi angulato-costati, subglabri. [Annotinous branches terete, cinnamon-colored, densely hirsute and prickly; flower-bearing branchlets angular-costate, almost glabrous].

Rami hornotini primum lucidi, rubro-virides, demum fuscati cum lenticellis minutis punctati. [Hornotinous branches at first glossy, red-green, later are darkened, punctate with minute lenticels].

Ramuli novelli dense cinereo-tomentosi, veteri glabri, rubicundi vel castanei; gemmae ovato-semiconicae, acutae, costatae. [Young branchlets densely ash-grey-tomentose, old ones glabrous, ruddy or chestnut-brown: buds ovate-half conic, acute, ribbed].

C. Leaves in woody plants are described in the same way as in herbaceous plants. The only difference is in that in trees and shrubs there is more often made distinction between old and young leaves and among the leaves borne on the different kinds of shoots. These different types of leaves are described separately. Examples :

1. Folia novella sericea, adulta utrinque glabra, subtus pallidora. [Young leaves sericeous, mature ones on both surfaces glabrous, underneath more pale].

Folia novella albo-argentea vel sericea, adulta utrinque viridia, omnino glaberrima. [Young leaves silvery-white or sericeous, mature ones on both surfaces green, altogether very glabrous].

2. Folia ramulorum fertilium inferiora obovata, apice serrata vel integra, superiora obovata, profunde 5-7-incisa; folia ramorum

sterilium profunde partita. [Lower leaves of the fertile branch-lets obovate on the apex entire or serrate, upper leaves obovate, deeply 5-7-incised; leaves of the sterile branches deeply parted].

Folia turionum stipulata, lineari-lanceolata, folia ramorum flori-ferorum exstipulata, lanceolata vel oblonga. [Leaves of the turions stipulate, linear-lanceolate, leaves of the flowering branches exsti-pulate, lanceolate or oblong].

3. In all other respects, woody plants are described same as herbaceous ones.

4. Below are given examples of somewhat standardized diag-noses of hypothetic woody species belonging to different systematic groups.

1. *Palm.*

Name. Authority sp. nov.

Palma elata, ad...m. alta. Caudex annulatus, aculeatus. Folia ca....m. longa, pinnata. Rachis aculeata utrinque cum segmentis numero...Spadix ca....m. longus, ramosus. Flores masculini in apicibus ramorum. Sepala minuta. Corolla rubescens, profunde divisa. Stamina 6. Filamenta complanata. Antherae oblongae. Flores feminaei subglobosi. Sepala minuta, reniformia. Petala rigida. Ovarium subglobosusm. Fructus glaber, depressoglobosus. Semen depresso-globosum.

[Tall palm up to...m. high. Trunk marked with rings, prickly. Leaves ca...m. long, pinnate. Rachis prickly on both sides with... segments. Spadix ca....m. long, branched. Male flowers are on the tips of the branches. Sepals minute. Corolla turning red, deeply divided. Stamens 6. Filaments flattened out. Anthers oblong. Female flowers subglobose. Sepals minute, kidney-shaped. Petals rigid. Ovary subglobose. Fruit glabrous, depressed-globose. Seed depressed-globose].

Hab....................

Area geogr............................

Affinis............, sed differt ....

Specimina examinata: ....

2. *Tree (Dicot.)*

Name. Authority sp. nov.

Arbor magna, elata, ca....m. alta. Truncus ca....cm. diametro. Cortex griseus, varie fissus. Rami et ramuli cinereo-fusci, glaberrimi. Gemmae parvae, ovatae. Folia elliptica vel obovata...cm. longa,...cm. lata, basi truncata, apice acuminata. Flores in inflorescentiis terminalibus. Perigonium... -fidum, lobis ovatis. Stamina cum filamentis filiformibus. Ovarium subglobosum, pilosum, styli... -fidi. Fructus hirtellus, ...cm. longus, ...cm. crassus.

[Large, tall tree, ca....m. high. Trunk ca....cm. across. Bark gray, variously fissured. Branches and branchlets ash-grey-somber brown, completely glabrous; Buds small, ovate. Leaves elliptic or obovate...cm. long,...cm. broad, truncate at base, acuminate at the apex. Flowers in terminal inflorescences. Perigon divided in... parts, with ovate lobes. Stamens with thread-like filaments. Ovary subglobose, pilose, styles divided in...parts. Fruit minutely hairy...... cm. long.......cm. thick].

Hab.................

Area geogr.............................

Affinis............, sed differt....................

Specimina examinata :..........................

3. *Shrub (erect).*

Name. Authority sp. nov.

Frutex erectus, ramis rubro-fuscis. Ramuli graciles, primum adpresse piloselli. Gemmae minutae, ovatae. Folia ovata vel oblongo-ovata, basi subtruncata, apice acuta,...cm. longa,...cm. lata. Inflorescentiae axillares. Perigonium... -fidum, lobis obovatis. Stamina parva, flavida. Ovarium ellipsoideum, styli filiformes. Fructus verrucosi, ...cm. longi, ...cm. crassi.

[Erect shrub with red-somber brown branches. Branchlets slender, at first adpressed minutely pilose. Buds small, ovate. Leaves

ovate or oblong-ovate, almost truncate at base, acute at the apex,...
cm. long,...cm. broad. Inflorescences axillary. Perigon divided
in...parts, with obovate lobes. Stamens small, yellowish. Ovary
ellipsoidal, styles thread-like. Fruits warty,......cm. long,......cm.
thick].

Hab... ... ...

Area geogr... ... ...

Affinis........., sed differt... ... ...

Specimina examinata: ... ...

    4. *Scandent shrub (or liane).*

Name   Authority sp. nov.

Frutex alte scandens, ca...m. longus. Cortex fuscescens, sub-
erosus. Rami et ramuli fusci, novelli puberuli, adulti glabri.
Gemmae ovatae, velutinae. Folia rotundato-subcordata,...cm. longa,
...cm. lata, integerrima, utrinque subglabra. Flores axillares,...mm.
longi. Pedicelli filiformes, bracteati. Calyx puberulus, cyathiformis.
Corolla patens,...... -petala. Stamina cum filamentis filiformibus.
Antherae subglobosae. Ovarium ovoideum, glabrum. Stylus crassus,
apice ...-fidus. Fructus oblongus, pendulus, ...cm. longus,...cm.
crassus.

[High climbing shrub, ca...m. long. Bark turning somber brown,
corky. Branches and branchlets somber brown, young downy with
very short soft hairs, mature glabrous. Buds ovate, velvety.
Leaves rotundate-subcordate, ...cm. long,...cm. broad, completely
entire, on the both surfaces almost glabrous. Flowers axillary,...
mm. long. Pedicels thread-like, bracteate. Calyx downy with
very short soft hairs, cup-shaped. Corolla spreading with...petals.
Stamens with thread-like filaments. Anthers subglobose. Ovary
ovoid, glabrous. Style thick, on the apex divided into...parts.
Fruit oblong, pendulous, ...cm. long, ...cm. thick].

Hab.... ... ...

Area geogr... ... ...

Affinis... ... ..., sed differt......

Specimina examinata: ... ...

IV. Description of infraspecific taxa.

1. Descriptions of infraspecific taxa are usually simpler and shorter than descriptions of species. There are usually less characters involved in the description of infraspecific taxa.

2. The Latin terms for the designation of new infraspecific taxa are as follows:

   New subspecies = subspecies nova = subsp. or ssp. nov.

   New variety = varietas nova = v. or var. nov.

   New form = forma nova = f. or fo. nov.

3. The head line in the description of a variety (or any other infraspecific taxon) may be presented in two forms: (1) with synonymy and (2) without synonymy.

   Examples:

   1. *Dryopteris laeta* Christ Ind. Fil. (1905) 273; Fomin in Fl. Sib. Orient. Extr. V (1930) 74.

      Nephrodium laetum Kom. Fl. Mansh. I (1901)

      var. oblongifrons Kitag. var. nov.

   2. *Axyris amaranthoides* L. Sp. Pl. (1753) 979

      var. dentata var. nov.

   Note : The second variant of the head line is more usual.

4. The wording in the body of the diagnosis of a variety may be varied in different ways. To illustrate this fact, below are given five variants of one and the same description of a hypothetical variety.

   Examples:

   1. Foliis latioribus, pilosis a typo distincta. [It is distinct from the typical plants in more broad, pilose leaves].

   2. Foliis latioribus, pilosis a typo recedit. [It deviates from the typical plants in more broad, pilose leaves].

   3. Foliis latioribus, pilosis differt (distincta or recedit.) [It is different (distinct or deviates) in more broad, pilose leaves].

   4. Variat foliis latioribus, pilosis. [It varies in more broad,

pilose leaves].

5. Folia latiora, pilosa. [Leaves more broad and pilose].

These examples show that the diagnosis of a variety consists of enumeration of distinctive characters of this variety, with addition of one of the following words : distincta=distinct, differs=it differs, recedit=it deviates, and variat=it varies. But, sometimes, in order to make diagnosis more short, these words are omitted (e.g. in example 5).

5. Sometimes to the diagnosis is added the concluding phrase : "caeterum ut in typo"=in other respects is like the type. In this case the diagnosis assumes the following form:

Foliis latioribus, pilosis differt. Caeterum ut in typo. [It differs in more broad, pilose leaves. In other respects is like the type].

Note : The expression "caeterum ut in typo" *can not* be used if the diagnosis ends in the words: "a typo recedit" (or distincta, or differt, etc.)

6. In the end of the diagnosis are necessary, as usual, indications of the ecology, geographical distribution and type specimens of the new variety. Sometimes, instead of the indication of ecology and geographical distribution, is used expression: "habitat cum typo" which means: grows with the typical plants.

7. The diagnoses of other infraspecific taxa are written in the same way as the diagnosis of a variety.

§3.   Variation of the form of diagnosis and of its parts.

1. Diagnoses of the species, with regard to their completeness, may be divided into complete and incomplete. The difference between these types is illustrated by the following examples:
Examples :

1.   Complete type.

Name Authority sp. nov.

Planta perennis. Radix crassa, lignosa. Caulis strictus, rube-

scens,...cm. altus, ramosus. Folia ovata, petiolata, acuta, integra, ...
cm. longa, ...cm. lata, utrinque glabra. Flores albi, ca...cm.
diametro. Pedicelli filiformes, bracteati. **Calyx glaber.** Corolla
patens, etc. Fructus oblongus, glaber, viridis, ...cm. longus, ...cm.
crassus.

Habitat... ... ...

Area geographica... ... ...

Speciei... ...   affinis, sed... ...   differt.

Typus speciei: ... in herbario   ... ... conservatur.

    2. Incomplete type.

Name Authority sp. nov.

Speciei...affinis, sed... ... ... differt.

Typus speciei :... ... ... in herbario... ... conservatur.

These examples show that incomplete type differs from the complete in the absence of morphological descriptions and of descriptions of ecology and geographical distribution of a new
species. The incomplete type, thus, consists only of the diagnosis of species in the narrow sense. In descriptive botany
both types of diagnoses are used, but the incomplete one is used
much less frequently.

2. The diagnosis (in the narrow sense) of a new species, i. e. the
discussion of its relationships may be placed in two different
ways: (1) immediately after the body of the description; or (2)
right before the latter.

Examples:

    1. Diagnosis (marked with the asterisk) is placed after the body
of the description.

      Majanthemum intermedium Worosch. nov. spec.

Body of the description

	Differt a *M. dilatato* (How.) Nels. et Macbr. foliis minoribus, etc.
*	

Hab. in silvis.

Typus...   ...   ...

2. Diagnosis (marked with the asterisk) is placed before the body of the description.

Cirsium kagomontanum Nakai sp. nov.

*
> Affine *C. schantarense* sed ex quo differt capite multo minore caule elatiore, pedunculo gracile et ramoso.

> Body of the description

Hab. Nippon, etc.

3. Some authors publishing new plant names use, to show their responsibility for these names, not their own name, but: (1) the word *"mihi"* (abbreviated m.), which means: to me; or (2) the word *"nobis"* (abbreviated nob.), which means: to us; or (3) do not put anything at all.

Note : the word *mihi* is used when there is only one author of the name and the word *nobis* when there are several authors of the name. Both of these words are put behind the new taxon's name as is shown by following examples.

Examples :

(1) Callicarpa pilosissima *mihi* (or *m.*) spec. nov.

(2) Callicarpa pilosissma *nobis* (or *nob.*) spec. nov.

(3) Callicarpa pilosissima spec. nov.

The way of designation authority by the means of words *mihi* and *nobis* was usual with the early authors but is not much practiced today.

4. The descriptions of plants may be also divided into two types with regard to the case in which they are written. In the first category fall descriptions written in the Nominative case and in the second those written in Ablative case. As is well known

the last type was traditional for the classical botanical writings. The characteristic feature of these classical descriptions is that almost all words in them are in Ablative case and that often they consist of only one sentence. This is because all words in these descriptions are modified with preposition *cum* which is, however, omitted in the text.

The difference between these types of descriptions is illustrated by the following examples.

Examples:

1. Description of an hypothetical species written in Nominative case. Perennis. Radix valida. Caulis erectus, glaber, ca. 45 cm. altus. Rami numerosi, erecti. Folia sessilia, oblonga, 2-3 cm. longa, 0,7-1,5 cm. lata, utrinque glabra, margine integra. Flores axillares, 5 mm. longi, 3 mm. diametro. Pedicelli puberuli, bracteati. Calyx angusuts, pilosus. Corolla rosea, patens. Fructus ovoideus, viridis, pendulus.

2. Same description written in Ablative case.

Perennis, radice valida, cauli erecto, glabro, ca. 45 cm. alto, ramis numerosis, erectis, foliis sessilibus, oblongis, 2-3 cm. longis, 0,7-1,5 cm. latis, utrinque glabris, margine integris, floribus axillaribus, 5 mm. longis, 3 mm. diametro, pedicellis puberulis, bracteatis, calyce angusto, piloso, corolla rosea, patenti, fructu ovoideo, viridi, pendulo.

Although, today, diagnoses of species are sometimes written in Ablative case, in imitation to the classical authors, it is better and more correct to write them in Nominative case. This method of writing descriptions is recommended in this guide. The descriptions written in Nominative case are more clear and easy to understand and it is more consistent to write descriptions of species in Nominative case since descriptions of taxa of all other ranks are written in this case. Besides, the method of writing descriptions of species in Ablative case is out of date.

Some contemporary authors imitating early Latin writings, do not use any punctuation in their plant descriptions, except for the fullstop (period). This manner of writing descriptions can not be recommended, because such descriptions are difficult to understand and this may create confusion.

4. Dedications.

1. In many cases taxa are named in honor of some person.
2. The dedication of a taxon to this person usually ends the botanical description and is written in the following way:
   1. Clarissimi(ae)......(1)......hanc genus (or speciem, varietatem, etc.) dedicavi. [I have dedicated this genus (or species, variety, etc.) to......]
   2. Genus (or species, varietas, etc.) meum(mea) clarissimi(ae)... (1)......dedico. [I dedicate my genus (species, variety, etc.) to......]
   3. Nomen generis (or speciei, varietatis, etc.) in honorem clarissimi(ae)......(1)......datum est. [The name of the genus (or species, variety, etc.) is given in honor of......]
   4. Genus (or species, varietas, etc.) in memoriam clarissimi...... (1)......dedicavi (or nominavi). [I have dedicated (or I have named) the genus (or species, variety, etc.) to the memory (or in memory) of......]

The above models are the formulae for writing dedications. The versions 1-3 are used when the person, in whose honor is named the taxon, is still living. The version 4 is used when this person is dead. The place for the name of this person in the formulae is shown by arabic numeral 1 in parentheses. In versions 1 and 2 this name must be put in Dative case and in versions 3 and 4 in Genitive case. Besides, the word *"clarissimi"* is used in above formulae if the person in whose honor is named the taxon is of male sex and the word *"clarissimae"* is used if this person is of female sex. In version 2 the word *genus* is of the neuter gender therefore to it corresponds neuter pronoun *meum*; the words: *species, subspecies, varietas,* and *forma* are all of the feminine gender, therefore to them corresponds the feminine pronoun *mea.*

Examples of dedications :

Centaurea Modesti Fed.

Cl. Prof. doct. M. M. Iljin nomen speciei dedico.

Acantholimon Alexandri Fed.

Nomen speciei datum in honorem fratri mei Alexandri, qui iconibus formosis opusculum meum ornavit.

Carex Slobodovii V. Krecz.

Nomen in honorem A. A. Slobodov florae Asiae Mediae investigatoris diligentissimi.

Carex Kabanovii V. Krecz.

Nomen in honorem cl. N. E. Kabanov florae sachalinensis investigatoris datum est.

Viola Elisabethae Klokov

Species mea cl. Elisabethae Steinberg dedico.

Nepeta Schischkinii Pojark.

Hanc speciem in honorem cl. B. Schischkin florae URSS investigatoris diligentissimi dedico.

Nepeta Fedtschenkoi Pojark.

Species nova in honorem cl. B. Fedtschenko florae turkestanicae exploratoris diligentissimi nominata.

Phlomis Drobovii M. Pop.

Ad honorem collectoris nominata est.

Schnabelia Hand.- zt.

Genus hoc dom. R. Schnabel, in urbe Tschangscha negotiateri, gratiam reddens dedico.

Since today few people put every part of the description in Latin, I give below, for the guidance of the readers, several examples of dedications in English.

Emmenopterys Rehderi Metcalf

It gives me great pleasure to name this in honor of Mr. **Alfred Rehder**, Curator of the Arnold Arboretum, with whom I have been closely associated for the last two years and whose kindly suggestions and advice have been a great help and inspiration.

Chelone Scouleri Douglas

This species was named by Mr. Douglas in honor of Dr. Scouler, the companion of his voyage to the west coast of America, who was, we understand, been recently appointed to the chair of Natural History in the University of Glasgow.

Prunus Twymaniana Koehne

At the request of Mr. Wilson I have named this species in compliment to Bertie Twyman, Esq., of the British Consular service in China, who was of very considerable assistance to Mr. Wilson during 1908.

Potamogeton Drucei Fryer

This plant was first found by Mr. G. C. Druce in the river Loddon, Berkshire, August, 1893, and I have great pleasure in dedicating it to the earnest worker who has done so much to advance British topographical Botany, and who at once recognised the distinctness of the plant from others of the genus.

Deutzia Schneideriana Rehder

I take pleasure in associating with this species the name of Mr. C. K. Schneider whose "Beitrag zur Kenntniss der Gattung Deutzia" is a valuable contribution to the knowledge of this genus.

Ilex Wangiana

This species is named after the collector, a fellow-student at the Arnold Arboretum, Mr. C. W. Wang.

Rosa Helenae Rehder et Wilson

This new species is named for my wife (E. H. Wilson)

Prunus Conradinae Koehne

This species is named for the wife of the author.

SECTION II.   THE TAXONOMIC CITATION.

§ 1.   Citation of Authors' Names.

In accordance with the International Rules of Botanical Nomenclature taxonomic citation of Authors' names should be made in following way :

1. Authors' names put after names of plants should be abbreviated unless they are very short. For this purpose, particles are omitted unless they are an inseparable part of the name, and the first letters are given without any omission.

   Examples : *Lam.* for J. B. P. A. Monet chevalier de Lamarck; but *De Wild.* for E. De Wildeman; *Beauv.* for A. M. F. J. Palisot de Beauvois; *D'Urv.* for Dumont D'Urville J. S. C.; but *Du Roi* for J. P. Du Roi.

2. If a name of one syllable is long enough to make it worthwhile to abridge it, the first consonants only are given.

   Example : *Fr.* for Elias Magnus Fries.

3. If the name has two or more syllables, the first syllable and the first letter of the following one are taken, or the two first when both are consonants.

   Example : *Juss.* for Jussieu; *Rich.* for Richard; *Pall.* for Pallas.

4. When it is necessary to give more of a name to avoid confusion between names beginning with the same syllable, the same system is to be followed. For instance, two syllables are given together with the one or two first consonants of the third; or one of the last characteristic consonants of the name is added.

   Examples : *Bertol.* for Bertoloni, to distinguish it from Bertero, *Michx.* for Michaux, to distinguish it from Micheli; *Wangh.* for Wangenheim, to distinguish it from Wangerin.

5. Christian names or accessory designations serving to distinguish two botanists of the same name are abridged in the same way.

Examples : *Adr. Juss.* for Adrien de Jussieu; *Gaertner f.* for Gaertner the son; *R. Br.* for Robert Brown; A. *Br.* for Alexander Braun; *J. F. Gmel.* for Johann Friedrich Gmelin; *J. G. Gmel.* for Johann Georg Gmelin; *C. C. Gmel.* for Carl Christian Gmelin; *S. G. Gmel.* for Samuel Gottlieb Gmelin; *Schulz Bip.* for Karl Heinrich Schulz (Bipontinus, i.e. of Zweibruecken); *Muell. Arg.* for Jean Mueller (Argoviensis, i.e. of Aargau), to distinguish it from P. J. Mueller; *L. f.* for Linnaeus the son.

6. In several cases the names are abridged, according to a custom, in different manner. International Code recommends to conform to such well-established custom.

Examples : *L.* for Linnaeus; *DC.* for De Candolle; *St.-Hil.* for Saint Hilaire; *F.v.Muell.* for Ferdinand von Mueller; *H.B.K* for Humboldt, Bonpland and Kunth; *M. B.* for Marschall von Bieberstein; *C. A. M.* for Carl Anton Meyer; *Ldb.* for Carl Friedrich von Ledebour; *W. et K.* for Waldstein et Kitaibel; *Bge.* for Alexander von Bunge; *Rgl.* for E. A. von Regel; *B. S. P.* for Britton, Sterns and Poggenberg.

7. When a plant name has been published jointly by two authors, the names of both should be cited, linked by means of word *et* or by the sign "&".

Examples: *Didymopanax Gleasonii* Britton et Wilson (or Britton & Wilson);

Artemisia scoparia W. et K. (or W. & K.); *Lamium barbatum* Sieb. et Zucc. (or Sieb. & Zucc.)

8. When a plant name has been published jointly by more than two authors, the citation should be restricted to that of the first one followed by *et al.* Example : *Angelica hirsutiflora* Liu et al. for *Angelica hirsutiflora* Liu, Chao et Chang.

9. When a plant name has been proposed or given, but not published, by one botanist, and subsequently published by another, it is cited in a way shown by the following examples:

Examples: *Havetia flexilis* Spruce ex Planch. et Triana (proposed by Spruce and published by Planch. and Triana).— *Gossypium tomentosum* Nutt. ex Seem. (proposed by Nuttal and published by Seemann).—*Artemisia selengensis* Turcz. ex Bess. (proposed by Turczaninov and published by Besser).—*Carex Knorringiae* Kuek. ex B. Fedtsch. (proposed by Kuekenthal and published by B. Fedtschenko).

The same holds for names of garden origin cited as "Hort." (Hortulanorum).

Example: *Gesneria Donklarii* Hort. ex Hook.

Note: If it is desirable or necessary to abbreviate such a citation the name of the publishing author, being the more important, should be retained.

Examples: *Havetia flexilis* Planch. et Triana.—*Gossypium tomentosum* Seem.—*Artemisia selengensis* Bess.—*Carex Knorringiae* B. Fedtsch.—*Gesneria Donklarii* Hook.

10. When a plant name is published by a man writing in someone else's publication, it is indicated in a manner illustrated by the following examples.

Examples: *Viburnum ternatum* Rehder in Sargent, Trees and Shrubs 2:37 (1907) (name given by Rehder in Sargent's publication).— *Carex enervis* C. A. M. in Ldb. Fl. Alt. (name given by C. A. Meyer in Ledebour's Flora Altaica).—*Eleocharis mamillata* Lindb. fil. in Doerfler, Herb. Norm. (name given by Lindberg the son in Doerfler's publication).

Note: If it is desirable or necessary to abbreviate such a citation the name of the author who supplied the description is the most important and should be retained.

Examples: *Viburnum ternatum* Rehder.—*Carex enervis* C.A.M. —*Eleocharis mamillata* Lindb. fil.

11. When a species originally described in one genus, is later transferred to another genus, the name of the author of the

original specific epithet (if the species retains it) is placed in parentheses, and this is followed by the name of the author who has placed the species in the accepted genus.

Examples: *Adenophora stenanthina* (Ldb.) Kitagawa (described by Ledebour in the genus *Campanula* and transferred by Kitagawa into the genus *Adenophora*).—Chosenia macrolepis (Turcz.) Kom. (described by Turczaninov in the genus *Salix* and transferred by Komarov into the genus *Chosenia*).—*Fimbristylis verrucifera* (Maxim.) Makino (described by Maximovicz in the genus *Isolepis* and transferred by Makino into the genus *Fimbristylis*).

Note: This new combination of names should be cited in the publication where it is published in the following way :

Chosenia macrolepis (Turcz.) Kom. comb. nov.

The Latin abbreviation *comb. nov.* means: *combinatio nova* or new combination of name and epithet. This abbreviation must be appended to the new nomenclatural combination to show that it is newly published.

12. When a taxon originally described in one rank, is later transferred to another rank, the name of the author of the original epithet (if the taxon retains it) is placed in parentheses, and this is followed by the name of the author who has changed the rank.

Examples: *Solidago virga-aurea* L. subsp. *coreana* (Nakai) Kitagawa (this subspecies was described by Nakai in the rank of variety and raised by Kitagawa into the rank of subspecies).— *Cleistogenes Nakaii* Honda f. *purpurascens* (Honda) Kitagawa (this form was described by Honda in the rank of variety and was transferred to the rank of form by Kitagawa).

Note: The taxon with the changed rank must be cited at the time of publication in the following way :

Solidago virga-aurea L. subsp. *coreana* (Nakai) Kitagawa stat. nov.

The Latin abbreviation *stat. nov.* means: *status novus* or new rank

and must be appended to the name of a taxon with changed rank
to indicate the change of the rank.

§ 2. Taxonomic citation of titles.

1. In accordance with International Code of Botanical Nomenclature a reference to bibliographic sources in a botanical publication should consist of the following items, in the order in which
they are treated below (the order of the items is shown by the
numerals in parentheses).

2. (*1*) *Name of Author(s)*. In a citation appended to the name of a
taxon the name of the author should be abbreviated as recommended above.

3. After the name of a taxon, an unabbreviated author's name
should be separated from what follows by a comma; an abbreviated name needs no punctuation other than the period (full
stop) indicating abbreviation.
Examples: *Aristida* L. Sp. Pl. 82. 1753 *Aristida* Linnaeus, Sp.
Pl. 82. 1753. *Phragmites* Adans. Fam. Pl. 2 : 34 (1763). *Phragmites* Adanson, Fam. Pl. 2 : 34 (1763).

4. (*2*) *Title*. After the name of a taxon, the title of a book is
commonly abbreviated, and the title of an article in a serial
is commonly omitted. In a citation appended to the name of a
taxon, no punctuation should separate the title from what
follows other than a period (full stop) indicating abbreviation.
Examples: Hook. f. Fl. Brit. Ind.—G.F. Hoffm. Gen. Umbell.—
H. B. K. Nov. Gen Sp.—L. Sp. Pl.—Michx. Fl. Bor.—Am.—
DC. Prodr.—Kom. Fl. Mansh.

5. (*3*) *Name of a serial*. Principal words should be abbreviated to
the first syllable, with such additional letters or syllables as
may be necessary to avoid confusion; articles, prepositions, and
other particles (der, the, of, de, et, etc.) should be omitted except
when the omission might create confusion. The order of
words should be that which appears on the title-page. Unneces-

sary words, subtitles, and the like should be omitted.

To avoid confusion among publications having the same name or very similar names, the place of publication or other distinguishing data should be added in brackets.

No punctuation other than a period (full stop) indicating abbreviation should separate the name of the serial from what follows.

Examples: Ann. Sci. Nat.; not Ann. des Sci. Nat.—Am. Journ. Bot.; not Amer. Jour. Bot.—Bot. Jahrb.; not Engl. Bot. Jahrb.—Flora [Quito] (to distinguish it from the well-known "Flora" published in Jena).—Hedwigia, not Hedwig.—Gartenflora; not Gartenfl.

6. (4) *Edition and series.* If a book has appeared in more than one edition, these subsequent to the first should be designated by "ed. 2", "ed. 3", and so forth. If a serial has appeared in more than one series in which the numbers of volumes are repeated, these subsequent to the first should be designated by a roman capital numeral or by "ser.2", "ser.3", and so forth.

Examples: G.F. Hoffm. Gen. Umbell. ed. 2—L. Sp. Pl. ed. 2.—Mem. Am. Acad. 1I (or ser. 2) (Memoirs of the American Academy of Arts and Sciences, New Series); not Mem. Am. Acad. N.S.

7. (5) *Volume.* The volume should be shown by an arabic numeral; for greater clarity this should be printed in boldface type. When volumes are not numbered, the years on the title-pages may be used as volume numbers. The volume-number should always be separated from the numbers of pages and illustrations by a colon.

Examples: Michx. Fl. Bor. Amer. *1* : 48. 1803.—Vasey, Bull. Torrey Club *11* : 64. 1884.—Fourn. Mex. Pl. *2* : 29. 1886.

8. (6) *Part or issue.* If a volume consists of a separately paged parts, the number of the part should be inserted immediately after the volume number (and before the colon), either in

parentheses or as a superscript. For volumes which are continuously paged, the designation of parts serves no useful purpose and leads to typographical errors.

Example: *Symphioglossum* Turcz. Bull. Soc. Nat. Mosc. *21*[1]: 225. 1848 or *Symphioglossum* Turcz. Bull. Soc. Nat. Mosc. *21(1)*: 225. 1848.

9. *(7) Pages.* Pages are shown by arabic numerals, except those otherwise designated in the original. If several pages are cited, the numbers are separated by commas; or if more than two consecutive pages are cited, the first and the last are given, separated by a dash.

Examples: Bot. Jahrb. *79*: 523.—Fl. Mansh. ed. 2. *1*: 294, 295. —Bot. Mag. Tokyo *48* : 92-94. *f. 9.*

10. *(8) Illustrations.* Figures and plates, when it is desirable to refer to them, should be indicated by arabic numerals preceded by f. (figure) and pl. (plate) or t. (*tabula*) respectively; for greater clarity these should be printed in italic type.

11. *(9) Dates.* The year of publication should end the citation; or, in lists of works to which reference is made by author and date, it may be inserted between the author's name and the title of his work. If it is desirable to cite the exact date, day month, and year should be given in that order. The date (in either position) may be enclosed in parentheses.

Examples: *Anacampseros* Sims, Bot. Mag. *33* : pl. 1367. 1811.— *Monochaetum* Naud. Ann. Sci. Nat. III. *4* : 48. pl. 2. 1845.— *Kalmia* Linnaeus, Sp. Pl. *1* : 391. 1753.—*Hedysarum gremiale* Rollins, Rhodora *42* : 230 (1940)—*Artocarpus melinoxylus* Cagnep. Bull. Soc. Bot. Fr. *73* : 87. 1926.—*Carex stenantha* Franch et Sav. Enum. pl. Jap. *2* : 146. 1879.—*Prinsepia utilis* Royle, Ill. Bot. Himal. Mount. 206. *tab. 38. f. 1.* 1839.

SECTION III. THE ABSTRACT OF THE LATIN GRAMMAR

1. The Noun

§ 1. The gender

1. The noun in Latin has three genders:
   - (1) genus masculinum=masculine gender
   - (2) genus femininum (or femineum)=feminine gender
   - (3) genus neutrum=neuter gender.

   The abbreviations used commonly for these genders are: m., f., and n. respectively.

2. The gender in Latin is:
   - (a) assigned to the noun according to its sex;
   - (b) grammatical gender which may not agree with the sex.

3. The gender of nouns is recognised:
   - (a) according to their meaning;
   - (b) according to their ending.

(a) According to their meaning are of:
   - (1) masculine gender, the names of months, rivers and persons of male sex, even if grammatically (i. e. according to the ending) their gender should be feminine.

   Examples: September, Januarius, Aprilis, Tiberis (Tiber river), Rhenus (the Rhine), botanista (botanist), incola (inhabitant).

   - (2) Feminine gender, names of persons of female sex, of trees (important), and of countries, islands and cities ending in -us.

   Examples: mater (mother), populus (poplar), fagus (beech), Aegyptus (Egypt), Rhodus (Rhodes Island), Caucasus (the Caucasus).

   - (3) neuter gender, indeclinable nouns, such as: gummi (gum), cacao (cocoa), etc.

§ 2.   Declensions (general remarks)

1.   Latin nouns are declinable.
2.   Cases.   In Latin language there are six cases answering to the following questions:

1. Nominativus=Nominative case, abbreviation N., who? what?
2. Genitivus=Genitive case   ,,   G. whose?
3. Dativus=Dative case   ,,   D. to whom?
4. Accusativus=Accusative (or Objective) case, abbreviation Acc., whom? what?
5. Ablativus=Ablative case, abbreviation Abl., by, with or from whom?
6. Vocativus=Vocative case.   ,,   ,, V. Who? what?

In botanical Latin are used only first five of these cases, the Vocative being excepted.

3.   Number.   The nouns in Latin have two numbers:

1. Singularis=Singular, abbreviation S.
2. Pluralis=Plural, abbreviation P.

4.   All nouns consist of the *base* or *stem* which is non-changeable, and of the ending which is changeable.
5.   Nouns in Latin belong to five declensions as shown by the endings of the Genitive Singular:

Declension	1	2	3	4	5
Ending in Genit. Sing.	*-AE*	*-I*	*-IS*	*-US*	*-EI*

Examples:

Declension	Nominat. Sing.	Genit. Sing.
1	plant-*a*	plant-*ae*
2	ram-*us*	ram-*i*
	foli-*um*	foli-*i*
3	caul-*is*	caul-*is*
	gen-*us*	gener-*is*
4	fruct-*us*	fruct-*us*
5	speci-*es*	speci-*ei*

6. The cases of nouns are known by their case-endings. These change according to the type of declension, gender and number of a noun.

§3. First declension

1. To this declension belong nouns ending in Nominative Singular in -*a* and in Genitive Singular ending in -*ae*.
2. To this declension belong, e.g., names of countries and continents ending in -*a*, for example: Africa, America, Australia, Europa, India, China, Manshuria, Mongolia, Siberia, etc.
3. Plant names ending in -*a* in Nominative Singular and in -*ae* in Genitive Singular, for example, Malva, -ae; Campanula, -ae; Tilia, -ae; Betula, -ae; Spiraea, -ae; Rosa, -ae, etc., also belong to this declension.

   Important note : The nouns of Greek origin ending in -*ma* in Nominative Singular, e.g., stigma, stoma, rhizoma, etc., and generic plant names of the same origin and ending, like Phyteuma, *do not belong* to this declension. They have special declension discussed in §14.
4. All nouns of the first declension are feminine.

   Exception to the rule: agricola, -ae; anachoreta, -ae; nauta, -ae which are masculine.
5. The nouns of the first declension always have in Dative and Ablative of the Plural the same ending -*is*.
6. The table of case-endings.   Model   declension

	S.	P.	S.	P.
N.	-a	-ae	planta	plantae
G.	-ae	-arum	plantae	plantarum
D.	-ae	-is	plantae	plantis
Acc.	-am	-as	plantam	plantas
Abl.	-a	-is	planta	plantis

§4. Second declension

1. To this declension belong nouns ending in Nominative Singular

in: *-ir*, *-er*, *-us*, *-um*, and in Genitive Singular ending in *-i*.

2. The nouns ending in *-ir*, *-er*, *-us* are masculine, and those ending in *-um* are neuter.

Exception to the rule: The generic names of trees ending in *-us* in Nominative Singular and in *-i* in Genitive Singular are feminine, e. g., Fagus, -i; Populus, -i; Ulmus, -i, etc. This exception does not hold, however, for the names of herbaceous plants, e.g., Ranunculus, -i; Lupinus, -i, etc. which are, as usual, masculine.

3. Dative and Ablative of *all nouns* in the second declension have case-endings: *-o* in Singular and *-is* in Plural.

4. Nominative and Accusative case-forms of *neuter nouns* in Plural have ending *-a*.

5. The tables of case-endings.

	Masculine		Neuter	
	S.	P.	S.	P.
N.	-us, -er, -ir	-i	-um	-a
G.	-i	-orum	-i	-orum
D.	-o	-is	-o	-is
Acc.	-um	-os	-um	-a
Abl.	-o	-is	-o	-is

6. Model declension (masculine).

	Singular			Plural		
N.	hortus	liber	vir	horti	libri	viri
G.	horti	libri	viri	hortorum	librorum	virorum
D.	horto	libro	viro	hortis	libris	viris
Acc.	hortum	librum	virum	hortos	libros	viros
Abl.	horto	libro	viro	hortis	libris	viris

Model declension (neuter).

	Singular		Plural	
N.	arboretum	herbarium	arboreta	herbaria
G.	arboreti	herbarii	arboretorum	herbariorum

D.	arboreto	herbario	arboretis	herbariis
Acc.	arboretum	herbarium	arboreta	herbaria
Abl.	arboreto	herbario	arboretis	herbariis

§ 5. Third declension.

1. To this declension belong nouns with different endings in Nominative Singular, which depend upon the gender of these nouns.

2. The nouns of this declension are divided into three groups which have three sets of forms according to which they are declined. Each group has its own peculiarities, but they all share one common feature: the ending -*is* in the Genitive Singular. On the basis of this feature they are all grouped into one declension.

3. Dative and Ablative forms of Plural (*in all genders*) in this declension have the ending -*ibus*.

4. Nominative and Accusative forms of Plural in *masculine* and *feminine* nouns have identical ending -*es*.

5. In *neuter* nouns Nominative and Accusative forms of Plural have identical ending -*a*.

6. In the third declension it is necessary to know separately for each noun its Genitive Singular form, because in contrast with the first and second declensions this form mostly differs considerably from the Nominative Singular form.
Examples:

Nominative Singular form.	Genitive Singular form.
genus	generis
margo	marginis
apex	apicis

7. On the other hand, some nouns of the third declension do not change in the Genitive Singular.

Examples:

Nominative Singular form.	Genitive Singular form.
caulis	caulis
clavis	clavis
classis	classis

8. The groups into which nouns of the third declension are divided are as follows:

    (1) Consonant nouns.

    (2) Vowel (or i-) nouns.

    (3) Mixed nouns.

§ 6. Third declension (consonant nouns).

1. The declension of consonant nouns is the regular, or typical, form of the third declension, the other two forms being irregular.

2. All consonant nouns are imparisyllabic, i.e. have different number of syllables in the Nominative Singular and Genitive Singular forms.

Examples:

Nominative Singular form.	Genitive Singular form.
flos	flo-ris
cor-tex	cor-ti-cis
ra-dix	ra-di-cis
a-pex	a-pi-cis
des-crip-ti-o	des-crip-ti-o-nis
ex-pla-na-ti-o	ex-pla-na-ti-o-nis

3. In Genitive Plural all these nouns have the ending -um.

4. Consonant nouns are subdivided into following subgroups:

        (a) Masculine nouns.

        (b) Feminine nouns.

        (c) Neuter nouns.

5. The tables of case-endings for consonant nouns.

	Masculine		Feminine	
	Sing.	Plur.	Sing.	Plur.
N.	-or, -os -er, -ex	-es	-is, -es, -ix	-es
G.	-is	-um	-is	-um
D.	-i	-ibus	-i	-ibus
Acc.	-em	-es	-em	-es
Abl.	-e	-ibus	-e	-ibus

	Neuter	
	Sing.	Plur.
N.	-men, -ur, -us, -ut	-a
G.	-is	-um
D.	-i	-ibus
Acc.	Acc.=N.	-a
Abl.	-e	-ibus

§7. Third declension (consonant nouns, masculine).

1. The masculine nouns of the third declension have in Nominative Singular the endings *-or, -os, -er, -ex*.
   Examples:
   auctor, flos, character, cortex.

2. Irregular nouns:
   (a) The noun arbor, -oris = tree, should be masculine (ending -or), but it is feminine.
   (b) The following generic plant names ending in *-ex* should be of masculine gender, but they are feminine: Atriplex, -icis; Carex, -icis; Ilex, -icis; Rumex, -icis; Vitex, -icis.
   (c) The following generic plant names ending in *-er* should be masculine, but they are neuter: Acer, -eris; Piper, -eris; Papaver, -eris; Siler, -eris.
   (d) The following nouns ending in *-er* should be of masculine gender, but they are of neuter gender; iter, itineris = travel;

tuber, tuberis=tuber; suber, suberis=cork; ver, veris=
spring.

(e) The noun verticillaster=verticillaster [false-whorl] might be
declined in two ways*).

3. Model declension of masculine nouns.

Singular

N.	auctor	flos	character	cortex
G.	auctoris	floris	characteris	corticis
D.	auctori	flori	characteri	cortici
Acc.	auctorem	florem	characterem	corticem
Abl.	auctore	flore	charactere	cortice

Plural

N.	auctores	flores	characteres	cortices
G.	auctorum	florum	characterum	corticum
D.	auctoribus	floribus	characteribus	corticibus
Acc.	auctores	flores	characteres	cortices
Abl.	auctoribus	floribus	characteribus	corticibus

§ 8. Third declension (consonant nouns, feminine).

1. The feminine nouns of the third declension have in Nominative
Singular the endings -o, -s and -x, with exception of the nouns
ending in -os and -ex, because these are masculine.
Examples:
   sectio, varietas, radix.

2. The generic plant names ending in -is, -es, and -ix also belong
to this declension.
Examples: Abies, -etis; Iris, -idis; Larix, -icis.

3. Irregular nouns:
   a. The nouns margo, -inis and stolo, -onis (ending in -o) should

---

*In N. Zabinkova et M. Kirpicznikov, Lexicon Latino-Rossicum pro
botanicis Mosqua-Leningrad, 1957, p. 251, this word is given as a masculine
noun of third declension. Stearn, Botanical Latin, London, 1966, p. 544
gives it as a masculine noun of second declension.

be feminine, but they are masculine.

b. The generic name Senecio, -onis (ending *-o*) should be feminine, but it is masculine.

c. The nouns: caespes, -itis; pes, pedis; fascis, is; unguis, -is; canalis, -is; mons, -tis; fons, -tis; dens, -tis; calyx, -icis, (endings *-x*, *-es*, *-is*, and *-s* with preceding consonant) should be feminine, but they are masculine.

d. The noun os, oris = mouth (ending *-s*) should be feminine, but it is neuter.

e. The generic name Ribes, -is should be feminine, but it is neuter.

f. The nouns ending in *-us* in Nominative Singular and in *-ris* in Genitive Singular are not feminine, but neuter.

Examples: corpus, -oris; genus, -eris; tempus, -oris; latus, -eris.

4. Model declension of feminine nouns.

Singular

N.	sectio	varietas	radix
G.	sectionis	varietatis	radicis
D.	sectioni	varietati	radici
Acc.	sectionem	varietatem	radicem
Abl.	sectione	varietate	radice

Plural

N.	sectiones	varietates	radices
G.	sectionum	varietatum	radicum
D.	sectionibus	varietatibus	radicibus
Acc.	sectiones	varietates	radices
Abl.	sectionibus	varietatibus	radicibus

§ 9. Third declension (consonant nouns, neuter).

1. The neuter consonant nouns of the third declension have in Nominative Singular the endings *-men. -us. -ut.*

Examples: nomen, genus, caput.

2. Model declension of neuter nouns.

Singular

N.	nomen	genus	caput
G.	nominis	generis	capitis
D.	nomini	generi	capiti
Acc.	nomen	genus	caput
Abl.	nomine	genere	capite

Plural

N.	nomina	genera	capita
G.	nominum	generum	capitum
D.	nominibus	generibus	capitibus
Acc.	nomina	genera	capita
Abl.	nominibus	generibus	capitibus

3. The only difference in the declension of neuter nouns from that of masculine and feminine nouns is that in Nominative and Accusative Plural they have ending *-a* instead of *-es* as in the latter.

§ 10. Irregular forms of the third declension.

1. There are three groups of nouns in the third declension changing irregularly:

a. Parisyllabic nouns with the Genitive Singular endings *-es* and *-is*.

Examples: caulis, caulis; clavis, clavis; pubes, pubis.

b. Nouns with two consonants before the ending *-is* in Genitive Singular.

Examples: pars, partis; dens, dentis; mons, montis.

c. Imparisyllabic neuter nouns ending in Nominative Singular in *-e*, and *-ar* and in Genitive Singular in *-is*.

Examples: vegetabile, -is; calcar, calcaris.

Note: Into the last group falls also the generic name *Secale, -is*.

2. The nouns of a. and b. groups are of all three genders, but change according to the same pattern. For this reason this type of

declension is sometimes called mixed.

3. As compared with the regular type of the third declension these nouns have only one peculiarity: the ending *-ium* instead of *-um* in Genitive Plural.

4. The table of case-endings.

	Singular	Plural
N.	-es, -is,	-es
	-rs, -ns	
G.	-is	-ium
D.	-i	-ibus
Acc.	-em	-es
Abl.	-e	-ibus

5. Model declension (mixed).

	Singular			Plural		
	f.	f.	m.	f.	f.	m.
N.	clavis	pars	dens	claves	partes	dentes
G.	clavis	partis	dentis	clavium	partium	dentium
D.	clavi	parti	denti	clavibus	partibus	dentibus
Acc.	clavem	partem	dentem	claves	partes	dentes
Abl.	clave	parte	dente	clavibus	partibus	dentibus

6. The nouns of the c. group are also called the i- or vowel nouns, because the vowel *i* plays important role in formation of their case-endings.

7. The declension of the vowel nouns has following peculiarities in comparion with the regular type of third declension:

   a. They have the ending *-i* instead of *-e* in Ablative singular.

   b. They have the ending *-ia* instead of *-a* in Nominative and Accusative Plural.

   c. They have the ending *-ium* instead of *-um* in Genitive Plural.

8. The table of case-endings and model declension.

	Singular	Plural	Singular		Plural	
			n.	n.	n.	n.
N.	-e, -ar	-ia	vegetabile	calcar	vegetabilia	calcaria
G.	-is	-ium	vegetabilis	calcaris	vegetabilium	calcarium

	Singular	Plural	Singular		Plural	
			m.	m.	m.	m.
D.	-i	-ibus	vegetabili	calcari	vegetabilibus	calcaribus
Acc.=N.		-ia	vegetabile	calcar	vegetabilia	calcaria
Abl.	-i	-ibus	vegetabili	calcari	vegetabilibus	calcaribus

§ 11. Fourth declension

1. To this declension belong masculine nouns with the ending *-us* both in Nominative and Genitive Singular, and neuter nouns with the ending *-u* in Nominative and the ending *-us* in Genitive Singular.

   Examples: sensus, -us; sexus, -us; cornu, -us.

2. Irregular nouns:

   a. Generic name *Quercus, -us* which instead of being masculine is feminine (cf. also §§ 1 and 4).

   b. The noun tribus, -us which is also feminine instead of being masculine.

3. Nominative and Accusative Plural forms of masculine nouns in the fourth declension have the identical ending *-us*.

4. Dative and Ablative Plural forms of masculine nouns have the identical ending *-ibus*.

5. Nominative and Accusative forms of neuter nouns have identical endings, *-u* in Singular and *-ua* in Plural.

6. The nouns of the fourth declension ending in *-us* in Nominative Singular have in Dative and Ablative Plural two types of endings *-ibus* and *-ubus*. Hence it is equally rightful to say quercibus and quercubus, or tribibus and tribubus.

7. The table of case-endings.

	Singular		Plural	
	Masculine	*Neuter*	*Masculine*	*Neuter*
N.	-us	-u	-us	-ua
G.	-us	-us	-uum	-uum
D.	-ui	-u	-ibus	-ibus
Acc.	-um	-u	-us	-ua
Abl.	-u	-u	-ibus	-ibus

8. Model declension.

	Singular		Plural	
	Masculine	*Neuter*	*Masculine*	*Neuter*
N.	sexus	cornu	sexus	cornua
G.	sexus	cornus	sexuum	cornuum
D.	sexui	cornu	sexibus	cornibus
Acc.	sexum	cornu	sexus	cornua
Abl.	sexu	cornu	sexibus	cornibus

§ 12. Fifth declension.

1. To this declension belong feminine nouns with endings *-es* in Nominative Singular and *-ei* in Genitive Singular.

    Examples: crassities, crassitiei; series, seriei; species, speciei; superficies, superficiei.

2. In Nominative and Accusative Plural these nouns always have ending *-es*.

3. In Dative and Ablative Plural they always have ending *-ebus*.

4. The table of case-endings and model declension.

	Singular	Plural	Singular	Plural
N.	-es	-es	species	species
G.	-ei	-erum	speciei	specierum
D.	-ei	-ebus	speciei	speciebus
Acc.	-em	-es	speciem	species
Abl.	-e	-ebus	specie	speciebus

§ 13. General conclusions concerning the declension of nouns.

1. In each declension the Dative and Ablative forms have identical endings.

2. All neuter nouns in all declensions have Nominative and Accusative forms with identical endings in Singular and Plural respectively.

3. All neuter nouns in all declensions have Nominative and Accusative Plural forms in the endings of which the last letter is *-a* (*-a, -ia, -ua*).

4. All nouns (except neuter) in Accusative Singular have endings terminating in *-m* (*-am, -em, -im, -um*).
5. All nouns in Genitive Plural have endings terminating in *-um* (*-arum, orum, -erum, -ium, -um, -uum*).
6. All nouns (except neuter) in Accusative Plural have endings terminating in *-s* (*-as, -es, -os, -us*).
7. The regularities stated above are shown on the following table on page 88.

§ 14. Declension of nouns of Greek origin.

1. In botanical Latin there are many nouns (botanical terms) of Greek origin.
2. The nouns of Greek origin are declined according to three declensions.
3. To the first declension belong feminine nouns with the ending *-e* in the Nominative Singular and the ending *-es* in the Genitive Singular.
   Examples: botanice, -es; micropyle, -es.
4. The table of case-endings and model declension.

	Singular	Plural	Singular	Plural
N.	-e	-ae	botanice	botanicae
G.	-es	-arum	botanices	botanicarum
D.	-ae	-is	botanicae	botanicis
Acc.	-en	-as	botanicen	botanicas
Abl.	-e	-is	botanice	botanicis

5. To the second declension belong masculine and neuter nouns; masculine nouns with the ending *-os* in the Nominative Singular and the ending *-i* in the Genitive Singular; neuter nouns with the ending *-on* in the Nominative Singular and the ending *-i* in the Genitive Singular.
   Examples:
   carpos, -i (masculine)
   embryon, -i; phyton, -i (neuter).

Summarized table of case-endings in all declensions for nouns and adjectives.

Declens	Gender	Case type		Singular N.	G.	D.	Acc.	Abl.		Plural N.	G.	D.	Acc.	Abl.
I	f			-a	-ae	-ae	-am	-a		-ae	-arum	-is	-as	-is
II	m			-us, -er	-i	-o	-um	-o		-i	-orum	-is	-os	-is
II	n			-um	-i	-o	-um	-o		-a	-orum	-is	-a	-is
III	m	consonant		-or, -os, -er, -ex	-is	-i	-em	-e		-es	-um	-ibus	-es	-ibus
III	f	consonant		-is, -es, -ix	-is	-i	-em	-e		-es	-um	-ibus	-es	-ibus
III	n	consonant		-men, -us, -ur, -ut	-is	-i	Acc.=N.	-e		-a	-um	-ibus	-a	-ibus
III	m	mixed		-es, -is, -ns, -rs	-is	-i	-em	-e		-es	-ium	-ibus	-es	-ibus
III	f	mixed			-is	-i	-em	-e		-es	-ium	-ibus	-es	-ibus
III	n	mixed			-is	-i	Acc.=N.	-e		-a	-ium	-ibus	-a	-ibus
III	n	vowel		-e, -ar	-is	-i	Acc.=N.	-i		-ia	-ium	-ibus	-ia	-ibus
IV	m			-us	-us	-ui	-um	-u		-us	-uum	-ibus (-ubus)	-us	-ibus (-ubus)
IV	n			-u	-us	-u	-u	-u		-ua	-uum	-ibus	-ua	-ibus
V	f			-es	-ei	-ei	-em	-e		-es	-erum	-ebus	-es	-ebus

6. The endings *-os* and *-on* in these nouns may be substituted by Latin endings *-us* and *-um* respectively and in this case the nouns are declined as the second declension masculine and neuter Latin nouns

7. The table of case-endings.

	Singular		Plural	
	Masculine	*Neuter*	*Masculine*	*Neuter*
N.	-os	-on	-i	-a
G.	-i	-i	-orum	-orum
D.	-o	-o	-is	-is
Acc.	-um	-on	-os	-a
Abl.	-o	-o	-is	-is

8. Model declension.

	Singular		Plural	
	Masculine	*Neuter*	*Masculine*	*Neuter*
N.	carpos	phyton	carpi	phyta
G.	carpi	phyti	carporum	phytorum
D.	carpo	phyto	carpis	phytis
Acc.	carpum	phyton	carpos	phyta
Abl.	carpo	phyto	carpis	phytis

9. To the third declension belong feminine and neuter nouns; feminine nouns with the ending *-is* in the Nominative Singular and the ending *-is* (or *-eos*) in the Genitive Singular; neuter nouns with the ending *-ma* in the Nominative Singular and the ending *-atis* in the Genitive Singular.

Examples:

synopsis, -is; anthesis, -is; basis, -is (feminine)

systema, -matis; -stoma, -matis; rhizoma, -matis (neuter).

10. The table of case-endings.

	Singular		Plural	
	Feminine	*Neuter*	*Feminine*	*Neuter*
N.	-is	-ma	-es	-mata
G.	-is (-eos)	-matis	-ium	-matum
D.	-i	-mati	-ibus	-matibus (-matis)
Acc.	-in (-im)	-ma	-is	-mata
Abl.	-i	-mate	-ibus	-matibus (-matis)

11. Model declension.

	Singular		Plural	
	Feminine	*Neuter*	*Feminine*	*Neuter*
N.	synopsis	stoma	synopses	stomata
G.	synopsis (synopseos)	stomatis	synopsium	stomatum
D.	synopsi	stomati	synopsibus	stomatibus (stomatis)
Acc.	synopsin (synopsim)	stoma	synopsis	stomata
Abl.	synopsi	stomate	synopsibus	stomatibus (stomatis)

Appendix: The gender of some generic names of Greek origin. Generic names ending:

(a) In *-pogon*, e.g., *Andropogon; Cymbopogon, Centropogon* etc., and in *-codon*, e.g., *Platycodon, Siphocodon, Leptocodon* are of masculine gender.

(b) In *-geton* are of two genders, e.g., *Potamogeton* is masculine and *Aponogeton* is neuter. The Greek word geiton (geton) is of common gender, i.e. of masculine or feminine according to the circumstances.

(c) In *-carpos* (latinized carpus), e.g., *Physocarpus, Phlojodicarpus* are of masculine gender.

(d) In *-mecon*, e.g., *Hylomecon, Eomecon, Dendromecon* are of feminine gender.

(e) In *-ma*, e.g., *Phyteuma, Asyneuma,* or in *-on* (latinized -um), e.g., *Chamaenerion=Chamaenerium, Lysichiton=Lysichitum,* etc., are of neuter gender.

(For further details cf. Intern. Code of Bot. Nomenclature, Recommendation 75A).

2. The adjective.

§ 1. General remarks.

1. Like nouns, adjectives in Latin have three genders (masculine, feminine and neuter) and two numbers (Singular and Plural).

2. Latin adjectives are declinable, but they do not have special declensions; they change according to the same pattern as nouns (cf. § 13) except that there are no adjectives of the fourth and fifth declensions.

3. With regard to the type of declension the adjectives are divided into two groups:
   (1) Adjectives of the first and second declension;
   (2) Adjectives of the third declension.

§ 2. The first and second declensions of adjectives.

1. Adjectives of the first and second declensions have three types of endings; each of these types corresponds to one of the generic forms of these adjectives. The endings *-us* and *-er* correspond to the masculine form of an adjective; the ending *-a* for the feminine form and the ending *-um* for the neuter form.
   Examples:

Masculine	ruber	pilosus
Feminine	rubra	pilosa
Neuter	rubrum	pilosum

2. These adjectives are declined in the following way: The feminine form is declined according to the first declension and the masculine and the neuter forms according to the second declension.

3. Model declensions.
   (a) First declension (feminine form).

	Singular		Plural	
N.	pilosa	rubra	pilosae	rubrae
G.	pilosae	rubrae	pilosarum	rubrarum
D.	pilosae	rubrae	pilosis	rubris

	Singular		Plural	
Acc.	pilosam	rubram	pilosas	rubras
Abl.	pilosa	rubra	pilosis	rubris

(b) Second declension (masculine form).

	Singular		Plural	
N.	pilosus	ruber	pilosi	rubri
G.	pilosi	rubri	pilosorum	rubrorum
D.	piloso	rubro	pilosis	rubris
Acc.	pilosum	rubrum	pilosos	rubros
Abl.	piloso	rubro	pilosis	rubris

(c) Second declension (neuter form).

	Singular		Plural	
N.	pilosum	rubrum	pilosa	rubra
G.	pilosi	rubri	pilosorum	rubrorum
D.	piloso	rubro	pilosis	rubris
Acc.	pilosum	rubrum	pilosa	rubra
Abl.	piloso	rubro	pilosis	rubris

§ 3. The third declension of adjectives.

1. The adjectives of the third declension are divided into three groups according to the number of the generic endings which they have:

       (1) With one ending.
       (2) With two endings.
       (3) With three endings.

2. Adjectives of the first group have identical endings *-r*, *-x*, *-s* for all of the three generic forms, in Nominative Singular.

    Examples: par, paris; fugax, fugacis; simplex, simplicis; teres, teretis.

3. Adjectives of the second group have ending *-is* for masculine and feminine forms and the ending *-e* for the neuter form.

    Examples:

        Masculine and feminine vulgaris, viridis, tenuis.
        Neuter vulgare, viride, tenue.

4. Adjectives of the third group have ending -er for masculine form, ending -is for feminine form and ending -e for neuter form. There are only 12 of such adjectives in the Latin language and the five of them most useful for the plant taxonomist are listed below.

Masculine   acer,   puter,   silvester,   paluster,   terrester.
Feminine   acris,   putris,   silvestris,   palustris,   terrestris.
Neuter   acre,   putre,   silvestre,   palustre,   terrestre.

5. All adjectives of third declension *are declined as the vowel -or i-nouns* (cf. pp. 84, 85 of this guide). Consequently, they have in Ablative Singular ending -i instead of -e, in Genitive Plural ending -ium instead of -um, and for neuter adjectives ending -ia instead of -a in Nominative and Accusative Plural.

6. Irregular adjectives. As an exception to the rule, four adjectives of third declension are declined as consonant nouns. Of these is used in descriptive botany only the adjective impar, -aris.

7. Model declensions.

(a) *One-ending adjective.*

		Singular			Plural	
	m.	*f.*	*n.*	*m.*	*f.*	*n.*
N.	teres	teres	teres	teretes	teretes	teretia
G.	teretis	teretis	teretis	teretium	teretium	teretium
D.	tereti	tereti	tereti	teretibus	teretibus	teretibus
Acc.	teretem	teretem	teres	teretes	teretes	teretia
Abl.	tereti	tereti	tereti	teretibus	teretibus	teretibus

(b) *Two-endings adjective.*

		Singular			Plural	
	m.	*f.*	*n.*	*m.*	*f.*	*n.*
N.	viridis	viridis	viride	virides	virides	viridia
G.	viridis	viridis	viridis	viridium	viridium	viridium
D.	viridi	viridi	viridi	viridibus	viridibus	viridibus
Acc.	viridem	viridem	viride	virides	virides	viridia
Abl.	viridi	viridi	viridi	viridibus	viridibus	viridibus

(c) *Three-endings adjective.*

	Singular			Plural		
	m.	*f.*	*n.*	*m.*	*f.*	*n.*
N.	acer	acris	acre	acres	acres	acria
G.	acris	acris	acris	acrium	acrium	acrium
D.	acri	acri	acri	acribus	acribus	acribus
Acc.	acrem	acrem	acre	acres	acres	acria
Abl.	acri	acri	acri	acribus	acribus	acribus

§ 4. Degrees of comparison of adjectives

1. Qualitative adjectives in Latin have three degrees of comparison: positive, comparative and superlative.

2. Positive degree is the usual, unchanged, form of an adjective of the first, second or third declension.
   Examples: pilosus, -a, -um; viridis, -e; simplex, -icis.

3. The other two degrees are formed by adding of special endings to the stem of an adjective.

4. In order to find the stem of an adjective, it is necessary to take away, in Genitive Singular, the endings: *-ae* in the first declension; *-i* in the second declension, and *-is* in the third declension.
   Examples:

Gender	Nominative Sing.	Genitive Sing.	Stem
Masculine (2-nd decl.)	latus	lat (-i)	lat
Feminine (1-st decl.)	lata	lat (-ae)	lat
Neuter (2-nd decl.)	latum	lat (-i)	lat
Masc./Fem. (3-rd decl.)	viridis	virid (-is)	virid
Neuter (3-rd decl.)	viride	virid (-is)	virid
M./F./N. (3-rd decl.)	simplex	simplic (-is)	simplic

5. Comparative degree for all three declensions is formed by adding to the stem of an adjective the ending *-ior* for masculine and feminine forms and the ending *-ius* for neuter form.

Examples:
Gender

Masculine/Feminine	lat+ior=latior
Neuter	lat+ius=latius
Masculine/Feminine	virid+ior=viridior
Neuter	virid+ius=viridius
Masculine/Feminine	simplic+ior=simplicior
Neuter	simplic+ius=simplicius

6. Superlative degree is formed similarly to the comparative one, but the endings are: *-issimus* for the masculine form, *-issima* for the feminine form and *-issimum* for the neuter form.

Examples:
Gender

Masculine	lat+issimus=latissimus
Feminine	lat+issima=latissima
Neuter	lat+issimum=latissimum
Masculine	virid+issimus=viridissimus
Feminine	virid+issima=viridissima
Neuter	virid+issimum=viridissimum
Masculine	simplic+issimus=simplicissimus
Feminine	simplic+issima=simplicissima
Neuter	simplic+issimum=simplicissimum

7. Adjectives with irregular degrees of comparison.

(a). Adjectives in which masculine form ends in Nominative Singular in *-er* have comparative degree formed in regular way, but superlative degree formed by adding the endings: *-rimus* for masculine form, *-rima* for feminine form, and *-rimum* for neuter form. These endings are added to the *positive degree of masculine form of an adjective in Nominative Singular.*

Examples:

Gender	*Positive*	*Comparative*	*Superlative*
Masculine	ruber	rubr+ior	ruber+rimus

Gender	Positive	Comparative	Superlative
Feminine	rubra	rubr+ior	ruber+rima
Neuter	rubrum	rubr+ius	ruber+rimum
Masculine	acer	acr+ior	acer+rimus
Faminine	acris	acr+ior	acer+rima
Neuter	acre	acr+ius	acer+rimum

(b). Adjectives in which the masculine form ends in Nominative Singular in -*dicus*, -*ficus* and -*volus* have grades of comparison derived from an old grammatical form ending in -*ens* which is now out of use.

Examples:

Gender	Positive	Comparative	Superlative
Masculine	magnificus	magnificent +ior	magnificent +issimus
Feminine	magnifica	magnificent +ior	magnificent +issima
Neuter	magnificum	magnificent +ius	magnificent +issimum
Masculine	helvolus	helvolent +ior	helvolent +issimus
Feminine	helvola	helvolent +ior	helvolent +issima
Neuter	helvolum	helvolent +ius	helvolent +issimum

(c). Adjectives in which masculine form ends in Nominative Singular in -*us* with preceding vowel form comparative degree by adding the word *magis*=more and form superlative degree by adding the word *maxime*=most.

Examples:

Positive	Comparative	Superlative
arduus, -a, -um	magis arduus, -a, -um	maxime arduus, -a, -um
dubius, -a, -um	magis dubius, -a, -um	maxime dubius, -a, -um
idoneus, -a, -um	magis idoneus, -a, -um	maxime idoneus, -a, -um

(d). Six adjectives of third declension with masculine and feminine forms ending in Nominative Singular in *-ilis* form comparative degree in regular way, but superlative degree they form by adding endings *-limus, -lima* and *-limum* for the masculine, feminine and neuter forms respectively.

These endings are added to the stem of the adjective.

Examples:

Positive	Comparative	Superlative
facilis, -e	facilior, -ius	facillimus, -a, -um
difficilis, -e	difficilior, -ius	difficillimus, -a, -um
similis, -e	similior, -ius	simillimus, -a, -um
dissimilis, -e	dissimilior, -ius	dissimillimus, -a, -um
gracilis, -e	gracilior, -ius	gracillimus, -a, -um
humilis, -e	humilior, -ius	humillimus, -a, -um

(e). Adjectives with special degrees of comparison.

Positive	Comparative	Superlative
bonus, -a, -um	melior, -ius	optimus, -a, -um
malus, -a, -um	peior, -ius	pessimus, -a, -um
magnus, -a, -um	major, -ius	maximus, -a, -um
parvus, -a, -um	minor, minus	minimus, -a, -um
multus, -a, -um	plus, pluris	plurimus, -a, -um
propinquus, -a, -um	proprior, -ius	proximus, -a, -um
vetus, -eris	vetustior, -ius	veterrimus, -a, -um

(f). Artificially formed adjectives *which have no positive degree*. This degree is substituted in them by the comparative degree. In addition to that they have irregular superlative degree.

Comparative	Superlative
anterior, -ius	—
posterior, -ius	postremus, -a, -um
superior, -ius	supremus, -a, um or summus, -a, -um
inferior, -ius	infimus, -a, -um or imus, -a, -um
exterior, -ius	extremus, -a, -um
interior, -ius	intimus, -a, -um
prior, -ius	primus, -a, -um

Note: The word anterior stems from the adjective ante=before, earlier; the words superior and inferior stem from the adjectives superus, -a, -um=upper and inferus, -a, -um=lower respectively; exterior, -ius and posterior, -ius are comparative degrees of the adjectives exterus, -a, -um and posterus, -a, -um.

§ 5. Declension of degrees of comparison.

1. Adjectives being raised to the comparative and superlative degrees at the same time change their declension as follows:

   (a). Adjectives which in positive degree are of the *first and second declensions* are converted to adjectives of the *third declension* by raising to *comparative* degree. They are re-converted to adjectives of the *first and second declensions* by raising to *superlative* degree.

   (b). Adjectives which in positive degree are of the third declension remain in the same declension when raised to comparative degree, but they are converted to adjectives of the first and second declensions by raising to superlative degree.

2. The adjectives with endings -*ior*, -*ius* in comparative degree (third declension) have in Genitive Singular ending -*oris* with the accent on the letter o in penultimate syllable. They retain this accent also in all cases except Nominative Singular in masculine and feminine forms and except Accusative Singular in the neuter form.

3. Model declension.

Singular

	m.	*f.*	*n.*
N.	supérior	supérior	supérius
G.	superióris	superióris	superióris
D.	superióri	superióri	superióri
Acc.	superiórem	superiórem	supérius
Abl.	superióre	superióre	superióre

Plural

	m.	f.	n.
N.	superióres	superióres	superióra
G.	superiórum	superiórum	superiórum
D.	superióribus	superióribus	superióribus
Acc.	superióres	superióres	superióra
Abl.	superióribus	superióribus	superióribus

4. Superlative degree of adjectives with endings: *-issimus, -a, -um, -rimus, -a, -um,* and *limus, -a, -um* are declined according to the first and second declensions (cf. above).

5. Model declension.

	Singular			Plural		
	m.	f.	n.	m.	f.	n.
N.	summus	summa	summum	summi	summae	summa
G.	summi	summae	summi	summo-rum	summa-rum	summo-rum
D.	summo	summae	summo	summis	summis	summis
Acc.	summum	summam	summum	summos	summas	summa
Abl.	summo	summa	summo	summis	summis	summis

6. The adjective plus, pluris (comparative degree of multus, -a, -um) has irregular declension. In Singular it has no Dative and Ablative forms.

	Singular	Plural	
	m./f./n.	m./f.	n.
N.	plus	plures	plura
G.	pluris	plurium	plurium
D.	—	pluribus	pluribus
Acc.	pluris	plures	plura
Abl.	—	pluribus	pluribus

3. The verb.
§ 1. General remarks.

1. The three stems of the verb. There are following four funda-mental forms of Latin verbs given for each of them in dictiona-

ries: (1) The present (1-st person Singular, Present Tense, Indicative mood); (2) the Perfectum (1-st person Singular, Past Perfect, Indicative mood); (3) the Supine; (4) the Infinitive. From three of these forms (2,3,4) are derived: (1) the perfect stem of the verb, (2) the supine stem of the verb and (3) the infinitive stem of the verb which are basic for creation of its other forms.

2. Latin verbs have:
   (a) Two voices: Active and Passive.
   (b) Four moods: Indicative, Conjunctive, Imperative and Infinitive.
   (c) Six Tenses: Present, Past Imperfect, Past Perfect, Past Pluperfect, Future Tense I, and Future Tense II.
   (d) Two numbers: Singular and Plural.
   (e) Three persons.
   In this guide, however, are not discussed all forms of the verbs, but only those which are necessary for the work of taxonomist.

3. Latin verbs are divided into four conjugations.

4. Infinitive form of all Latin verbs ends in -re and by the letter which immediately precedes this ending might be known the type of conjugation of an verb, viz.:

Conjugation	The letter that precedes ending -re.
First	a
Second	$\bar{e}$ (long)
Third	$\breve{e}$ (short)
Fourth	i

Except for the third conjugation, these letters are the endings of the infinitive stems of the verbs (cf. § 2 item 2).

Examples of the verbs of four conjugations:

Conjugation	Verb
First	habito, habitare
Second	floreo, florēre
Third	lego, legĕre
Fourth	invenio, invenire

§ 2. Conjugation of verbs.

1. In order to make the personal forms of an verb in **Present Tense**, Indicative mood, it is necessary, first of all, to find the infinitive stem of this verb. This stem is found by putting away the ending *-re* in the infinitive form of all verbs except those belonging to the third conjugation. In the latter group of verbs the infinitive stem is found by putting away the ending *-ere* in the Infinitive form of an verb. This difference of the verbs of third conjugation depends upon the fact that the last letter of their stem is consonant and therefore the stem is joined with the ending with the help of connecting letter *e*.

2. The infinitive stems of the verbs are found in the following way:

Conjugations	Verb (Infinit. mood)	Verb stem	Stem ending
First	habita (-re)	habita	*-a*
Second	flore (-re)	flore	*-e*
Third	cresc (-ere)	cresc	*consonant*
Fourth	inveni (-re)	inveni	*-i*

Note: The ending of the verb's stem is always the same in all verbs of one conjugation.

3. After finding the stem of the verb, the following endings are joined to it in order to make the personal forms:
For the 1-st person Sing. *-o*; for the 1-st person Plur. *-mus*;
for the 2-nd person Sing. *-s*; for the 2-nd person Plur. *-tis*;
for the 3-rd person Sing. *-t*; for the 3-rd person Plur. *-nt*.

4. These endings are added directly to the stems of the verbs in the first and second conjugations, and in all persons, except the third person of Plural in the fourth conjugation.

5. In the third conjugation where the last letter of the stem is consonnant, the endings of the second and third persons of Singular and of the first and second persons of Plural are joined to the stem with help of connecting letter *i*. In the third person Plural both in the third and fourth conjugations the ending is joined to the stem with help of connecting letter *u*.

6. Model conjugations (Present tense, Indicative mood).

	I conjugation		*II conjugation*	
Person	Sing.	Plur.	Sing.	Plur.
1	habito	habitamus	florèo	florémus
2	habitas	habitatis	flóres	florétis
3	habitat	habitant	flóret	flórent

	III conjugation		*IV conjugation*	
Person	Sing.	Plur.	Sing.	Plur.
1	dívido	divídimus	invénio	invénimus
2	dívidis	divíditis	invénis	invénitis
3	dívidit	dividunt	invénit	invéniunt

7. For the description of plants it is necessary to know the Past Perfect forms of verbs (Indicative mood) which have following endings:

For the 1-st person Sing. -*i*; for the 1-st person Plur. -*imus*; for the 2-nd person Sing. -*isti*; for the 2-nd person Plur. -*istis*; for the 3-rd person Sing. -*it*; for the 3-rd person Plur. -*erunt*. These endings are joined to the Past Perfect stem of the verb.

8. The Past Perfect stem of an verb can be found by taking away the ending -*i* in the first person of Past perfect. This is done in the following way:

Conjugation	Past Perfect	Past Perfect stem
First	determinav (-i)	determinav
Second	floru (-i)	floru
Third	tex (-i)	tex
Fourth	inven (-i)	inven

9. Model conjugation (Past Perfect, Indicative mood).

	I conjugation		*II conjugation*	
Person	Sing.	Plur.	Sing.	Plur.
1	determinavi	determinavimus	florui	floruimus
2	determinavisti	determinavistis	floruisti	floruistis
3	determinavit	determinaverunt	floruit	floruerunt

	III conjugation		*IV conjugation*	
Person	Sing.	Plur.	Sing.	Plur.
1	texi	teximus	inveni	invenimus
2	texisti	texistis	invenisti	invenistis
3	texit	texerunt	invenit	invenerunt

1θ. Auxiliary verb sum = to be and its conjugation in Present tense, Indicative mood.

Person	Singular	Plural
1	sum	sumus
2	es	estis
3	est	sunt

4. The participle.

§ 1. General remarks.

1. Participle is an adjective derived from a verb.
2. In Latin there are following types of participles:
   (1) Of Present tense, Indicative mood, Active voice;
   Examples: fragrans, florens, scandens.
   (2) Of Past Perfect, Indicative mood, Passive voice;
   Examples: ornatus, doctus, instructus.
3. Participles of the first group are formed by adding to the infinitive stem of an verb the ending *-ns* for the verbs of the 1-st and 2-nd conjugations, and by adding the ending *-ens* for the verbs of the 3-rd and 4-th conjugations.
   Examples:

Conjugation	Verb	Infinit. stem	Participle
First	fragra (-re)	fragra	fragra+ns=fragrans
Second	florē (-re)	florē	flore+ns=flórens
Third	scand (-ĕre)	scand	scand+ens=scandens
Fourth	investi (-re)	investi	investi+ens=investiens

4. Participles of the second group are derived from the third fundamental form of the verb known as supinum. To the supinum stem are added endings: *-us* for the masculine, *-a* for the feminine

and *-um* for the neuter forms of the participle.

Examples:

Conjugation	Verb	Supinum	Supinum stem	Participle
First	ornare	ornat (-um)	ornat	ornat+us, -a, -um
Second	docēre	doct (-um)	doct	doct+us, -a, -um
Third	instruĕre	instruct (-um)	instruct	instruct+us, -a, -um
Fourth	invenire	invent (-um)	invent	invent+us, -a, -um

§ 2. Declension of participles.

1. Participles of the Present tense (1-st group) are declined in the same way as the adjectives of third declension with one ending.
2. Model declension.

	Singular m./f.	n.	Plural m./f.	n.
N.	scandens	scandens	scandentes	scandentia
G.	scandentis	scandentis	scandentium	scandentium
D.	scandenti	scandenti	scandentibus	scandentibus
Acc.	scandentem	scandens	scandentes	scandentia
Abl.	scandènti	scandenti	scandentibus	scandentibus

3. Participles of the Past Perfect tense (2-nd group) are declined in the same way as the adjectives of the first and second declensions.
4. Model declension.

	Singular m.	f.	n.	Plural m.	f.	n.
N.	tectus	tecta	tectum	tecti	tectae	tecta
G.	tecti	tectae	tecti	tectorum	tectarum	tectorum
D.	tecto	tectae	tecti	tectis	tectis	tectis
Acc.	tectum	tectam	tectum	tectos	tectas	tecta
Abl.	tecto	tecta	tecto	tectis	tectis	tectis

5. The preposition.

1. Latin prepositions are divided into three following groups:
   (1) Those requiring that the noun associated with them should be in the Accusative case.

Examples:

ad=to, up to, by, for

ante=before

circum=about, around

extra=outside

inter=between, among

intra=inside, within

per=through, on, by

post=after

prope=near

propter=because of

supra=above

trans=over, on the other side

versus=toward

(2) Those requiring that the noun associated with them should be in the Ablative case.

Examples:

a, ab=from

cum=with

de=down from

e, ex=out from

pro=for, instead of, as

sine=without

(3) Those which can be used with Accusative or Ablative case according to context. If preposition answers to the question when? or to where? (to what place?), i.e. indicates motion, the noun should be in Accusative case. If preposition answers to the question where? (in what place?), i.e. indicates fixed position, the noun should be in Ablative case.

There are only few such prepositions.

Examples:

in=in, on

sub=under

super=over

6. The conjunction.

1. Conjunctions are used for joining different members of the sentence.

2. The following 12 Latin conjunctions are commonly used in plant descriptions:

ac, atque, et, -que=and

aut, vel, sive, -ve=or

ut=as

autem, sed=but

tamen=however, nevertheless.

3. Of these conjunctions *-que* and *ve* are joined to the end of the words; conjunction *autem* is *never* put in the beginning, but always in the middle of the sentence and *after* the word which is associated with this conjunction.

7. The numeral.

§ 1. Cardinal and ordinal numerals.

1. In the contemporary botanical papers measurments are expressed in the metric system and all numbers are shown by arabic numerals. But in early botanical writings numbers were often shown by Roman numerals. Therefore below is given a list of the Roman cardinal numerals.

I=1=unus, -a, -um=one
II=2=duo, -ae, -o=two
III=3=tres, -es, ia=three
IV=4=quattuor=four
V=5=quinque=five
VI=6=sex=six
VII=7=septem=seven
VIII=8=octo=eight
IX=9=novem=nine
X=10=decem=ten
XI=11=undecim=eleven
XII=12=duodecim=twelve
XIII=13=tredecim=thirteen
XIV=14=quattuordecim= fourteen
XV=15=quindecim = fifteen
XVI=16=sedecim=sixteen
XVII=17 = septendecim= seventeen
XVIII=18=duodeviginti= eighteen

XIX=19=undeviginti=nine- teen
XX=20=viginti=twenty
XXI=21=unus et viginti= twenty one, etc.
XXVIII=28=duodetriginta = twenty eight
XXIX=29=undetriginta= twenty nine
XXX=30=triginta=thirty
XL=40=quadraginta=forty
L=50=quinquaginta=fifty
LX=60=sexaginta=sixty
LXX=70=septuaginta= seventy
LXXX=80=octoginta=eighty
XC=90=nonaginta=ninety
C=100=centum=one hundred
CC=200=ducenti, -ae, -a= two hundred

CCC=300=trecenti, -ae, -a= three hundred

CD=400=quadringenti, -ae, -a=four hundred

D=500=quingenti, -ae, -a= five hundred

DC=600=sescenti, -ae, -a= six hundred

DCC=700=septingenti, -ae, -a=seven hundred

DCCC=800=octingenti, -ae, -a=eight hundred

DCCCC or CM=900=non-genti, -ae, -a=nine hundred

M=1000=mille=one thousand

2 Of Latin cardinal numerals are declinable only unus, duo, tres, names of hundreds beginning from two hundred, and *milia* (plural of *mille*). The cardinals from quattuor (4) to centum (100) and mille (1000) are not changed. Of the declinable numerals unus is declined only in Singular and all the remaining only in Plural.

3. Declensions of numerals.

Singular

	m.	*f.*	*n.*
N.	unus	una	unum
G.	unius	unius	unius
D.	uni	uni	uni
Acc.	unum	unam	unum
Abl.	uno	una	uno

Plural

	m.	*f.*	*n.*	*m.*	*f.*	*n.*
N.	duo	duae	duo	tres	tres	tria
G.	duorum	duarum	duorum	trium	trium	trium
D.	duobus	duabus	duobus	tribus	tribus	tribus
Acc.	duos	duas	duo	tres	tres	tria
Abl.	duobus	duabus	duobus	tribus	tribus	tribus

	m.	*f.*	*n.*	*m.*	*f.*	*n.*
N.	trecenti	trecentae	trecenta	milia	milia	milia
G.	trecento-rum	trecenta-rum	trecento-rum	milium	milium	milium

	m.	f.	n.	m.	f.	n.
D.	trecentis	trecentis	trecentis	milibus	milibus	milibus
Acc.	trecentos	trecentas	trecenta	milia	milia	milia
Abl.	trecentis	trecentis	trecentis	milibus	milibus	milibus

Note: All names of hundreds from ducenti, -ae, -a through nongenti, -ae, -a are declined according to one pattern the example of which is given above.

4. Cardinal numerals in Latin are considered as adjectives. Hence those of them which are declinable should agree in gender, case and number with the noun with which they are associated.

Examples:

Unum folium; una spicula; unus fructus; tria segmenta; trium segmentorum; ex spiculis tribus una spicula mascula duae femineae.

5. Latin ordinal numerals, from one to ten, are listed below.

primus, -a, -um=first   sextus, -a, -um=sixth
secundus, -a, -um=second   septimus, -a, -um=seventh
tertius, -a, -um=third   octavus, -a, -um=eighth
quartus, -a, -um=fourth   nonus, -a, -um=ninth
quintus, -a, -um=fifth   decimus, -a, -um=tenth

6. Ordinal numerals in Latin are considered as the adjectives of the first and second declensions. Therefore they are declined as these and also therefore they should necessarily agree in gender, case and number with the nouns with which they are associated.

Examples:

Species prima; genus quartum; tribus decima; ordo secundus; spicula prima mascula, spiculae secunda et tertia femineae; genus primus a genere secundo structura florum differt.

§ 2. Distributive numerals and numeral-adverbs.

1. Distributive numerals answer the question in groups of how many?, e.g. bini=in pairs, two by two, etc.

2. Below are listed the first ten distributive numerals.

singuli, -ae = one by one	seni, -ae, -a = six by six
bini, -ae, -a = two by two	septeni, -ae, -a = seven by seven
terni, -ae, -a = three by three	octoni, -ae, -a = eight by eight
quaterni, -ae, -a = four by four	noveni, -ae, -a = nine by nine
quini, -ae, -a = five by five	deni, -ae, -a = ten by ten

3. Distributive numerals are declined in the same way as the adjectives of the first and second declensions, but *only in Plural*.

4. Numeral-adverbs answer the question how many times?, e. g. bis = twice, ter = thrice, etc. They are:

semel = once	sexies = six times
bis = twice	septies = seven times
ter = thrice	octies = eight times
quater = four times	novies = nine times
quinquies = five times	decies = ten times

Note: Numeral-adverbs are not declinable.

§ 3. Numerical prefixes.

1. The following prefixes are derived from the Latin numerals:

bi- = twice	multi- = many times
du- = twice	semi- = half
tri- = thrice	sesqui- = one and a half times
quadr- = four times	

2. These prefixes are used in forming of:

   (a) Adjectives, e.g. bipinnatus, -a, -um; trisectus, -a, -um; multifidus, -a, -um, etc.

   (b) Adverbs, e.g. duplo = twice as great, triplo = thrice as great, etc.

   (c) Multiplicative adjectives, e.g. simplex, -icis = single, etc.

3. The adjectives derived from numerical prefixes are declined in accordance with their endings as the adjectives of the first, second or third declension.

8. The pronoun.

1. Under the general name of pronoun in Latin are known posses-

sive, demonstrative and relative pronouns, and pronominal adjectives which all have three generic forms and all are declinable.

2. The use of pronouns and pronominal adjectives in plant descriptions is very limited. Therefore I will give here the examples of declension only of those of them which might be useful for the work of plant taxonomist.

3. Declensions of pronouns and pronominal adjectives.

   (a) Pronoun *ambo, ambae, ambo*=both is declined in the same way as the cardinal numeral duo, -ae, -o (cf. above).

   (b) Demonstrative pronouns: *hic, haec, hoc*=this; *ille, illa, illud*=that; *is, ea, id*=that, he, she, it; *idem, eadem, idem*= the same, the same as that previously mentioned, are declined in the following way:

	Singular			Plural		
	m.	*f.*	*n.*	*m.*	*f.*	*n.*
N.	hic	haéc	hoc	hi	hae	haec
G.	húius	húius	húius	hórum	hárum	hórum
D.	húic	húic	húic	his	his	his
Acc.	hunc	hanc	hoc	hos	has	haec
Abl.	hoc	hac	hoc	his	his	his

	Singular			Plural		
	m.	*f.*	*n.*	*m.*	*f.*	*n.*
N.	ílle	ílla	íllud	illi	íllae	ílla
G.	illíus	illius	illíus	illórum	illárum	illórum
D.	ílli	ílli	ílli	íllis	illis	íllis
Acc.	íllum	íllam	illud	illos	illas	ílla
Abl.	illo	ílla	íllo	íllis	íllis	íllis

	Singular			Plural		
	m.	*f.*	*n.*	*m.*	*f.*	*n.*
N.	is	éa	id	íi(éi)	éae	éa
G.	éius	éius	éius	eorum	earum	eorum
D.	éi	éi	éi	íis(éis)	íis(éis)	íis(éis)
Acc.	éum	éam	id	éos	éas	éa
Abl.	éo	éa	éo	íis(éis)	íis(éis)	íis(éis)

	Singular			Plural		
	m.	*f.*	*n.*	*m.*	*f.*	*n.*
N.	ídem	éadem	ídem	ídem (eídem)	eáédem	éadem
G.	eiúsdem	eiúsdem	eiúsdem	eorúndem	earúndem	eorúndem
D.	eídem	eídem	eídem	ísdem (eísdem)	ísdem (eísdem)	ísdem (eisdem)
Acc.	eúndem	eándem	ídem	eósdem	eásdem	éadem
Abl.	eódem	eádem	eódem	ísdem (eísdem)	ísdem (eísdem)	ísdem (eisdem)

(c) Relative pronouns *qui, quae, quod* = which (who) are declined in the following way:

	Singular			Plural		
	m.	*f.*	*n.*	*m.*	*f.*	*n.*
N.	qui	quaé	quod	qui	quae	quae
G.	cúius	cúius	cúius	quórum	quárum	quórum
D.	cui	cui	cui	quíbus	quibus	quibus
Acc.	quem	quam	quod	quos	quas	quae
Abl.	quo	qua	quo	quíbus	quíbus	quibus

(d) Pronominal adjectives *totus, -a, -um* = all and *nullus, -a, -um* = none are declined in the following way:

Singular

	m.	*f.*	*n.*	*m.*	*f.*	*n.*
N.	tótus	tóta	tótum	nullus	nulla	nullum
G.	totius	totíus	totíus	nullius	nullius	nullius
D.	tóti	tóti	tóti	nulli	nulli	nulli
Acc.	tótum	tótam	tótum	nullum	nullam	nullum
Abl.	tóto	tóta	tóto	nullo	nulla	nullo

Plural

	m.	*f.*	*n.*	*m.*	*f.*	*n.*
	m.	*f.*	*n.*	*m.*	*f.*	*n.*
N.	tóti	tótae	tóta	nulli	nullae	nulla
G.	totórum	totárum	totórum	nullorum	nullarum	nullorum
D.	tótis	tótis	tótis	nullis	nullis	nullis
Acc.	tótos	tótas	tóta	nullos	nullas	nulla
Abl.	tótis	tótis	tótis	nullis	nullis	nullis

9. The adverb.

§ 1. The nature of adverbs.

1. There are two kinds of Latin adverbs:
   (1) Primary adverbs (not derived from adjectives);
   (2) Derivative adverbs (derived from adjectives and participles).

2. Adverbs of the first group *have various endings.*
   Examples: subtus, demum, vix, jam.

3. Adverbs of the second group have endings *-e, -er, -iter, -im.*
   Examples: longe, tenuiter, repenter, radiatim.

4. When an adverb is formed the endings *-e, -er, -im, -iter* are joined to the adjective stems.

5. The ending *-e* is joined to the stems of the adjectives of *first and second declensions.* In order to find this stem it is necessary to take the masculine form of the adjective in Nominative Singular and to take away from this form the ending *-us.*
   Example of formation of adverb from the adjective of first and second declensions:

Adjective (masculine form)	Stem	Abverb
long (-us)	long	long+e=longe
lat (-us)	lat	lat+e=late
crass (-us)	crass	crass+e=crasse
alt (-us)	alt	alt+e=alte

6. If the masculine form of an adjective of the first and second declension is ending in *-tus,* the adverb from this adjective may be formed by adding ending *-im* to its stem.
   Examples:

Adjective (masculine form)	Stem	Adverb
radiat (-us)	radiat	radiat+im=radiatim
elongat (-us)	elongat	elongat+im=elongatim
annulat (-us)	annulat	annulat+im=annulatim

In other cases, however, the adverbs are formed from the adjectives of this type in regular way.

Examples:

Adjective (masculine form)	Stem	Adverb
lat (-us)	lat	lat + e = late
arct (-us)	arct	arct + e = arcte

7. The ending *-iter* is joined to the stems of the adjectives of the *third declension, with exception of the participles ending in -ns.*

8. In order to find the stem of an adjective of the third declension it is necessary to take this adjective in Genitive Singular and to take away from this form of adjective the ending *-is.*

Examples:

Adjective	Stem	Adverb
tenu (-is)	tenu	tenu + iter = tenuiter
acr (-is)	acr	acr + iter = acriter
vertical (-is)	vertical	vertical + iter = verticaliter

Note: As an exception the adverb with the ending -iter is formed from the adjective of the second declension *rarus*. The adverb thus formed is *rariter*. From the same adjective stems also the adverb *raro* (cf. below). Both adverbs have identical meaning.

9. From the participles ending in *-ns*, adverbs are formed by adding ending *-er* to the stems of these participles. The stems of the participles are found in the same way as those of the adjectives of the third declension.

Examples:

Participle	Stem	Adverb
repent (-is)	repent	repent + er = repenter
ascendent (-is)	ascendent	ascendent + er = ascendenter
scandent (-is)	scandent	scandent + er = scandenter

10. Many adjectives of the second declension in Ablative case Singular (ending *-o*) are used as adverbs.

Examples:

Adjective in Nominat. Sing.	Adjective in Ablative Sing.
creber	crebro
perpetuus	perpetuo
rarus	raro

Adjective in Nominat. Sing.	Adjective in Ablative Sing.
primus	primo
falsus	falso

The same holds for some nouns which in Ablative Singular are used as adverbs.

Examples:

Noun	Noun in Ablative Singular
fides, -ei	fide
sphalma, -matis	sphalmate
testis, -is	teste

§ 2. Grades of comparison of adverbs.

1. Adverbs derived from qualitative adjectives have grades of comparison.

2. With regard to the formation of grades of comparison Latin adverbs are divided into two groups:

   (1) Having regular grades of comparison.

   (2) Having irregular grades of comparison.

3. First group. Adverbs with regular grades of comparison.

The adverbs of this group have in comparative degree the ending -ius which is identical with the ending of the neuter form of respective adjective in comparative degree.

Examples:

Adjective	Adverb	
	Positive degree	Comparative degree
longus	longe	longius
tenuis	tenuiter	tenuius
acer	acriter	acrius

Superlative degree of the adverbs is formed by adding the ending -e to the stem of respective adjective raised in superlative degree. In order to find this stem it is necessary to take the masculine form of the adjective in Nominative Singular and to take from it away its ending -us.

Examples:

Adjective in positive degree	Adjective in superlat. degree	Stem	Adverb in superlative degree
longus	longissim (-us)	longissim	longissim+e=longissime
tenuis	tenuissim (-us)	tenuissim	tenuissim+e=tenuissime
acer	acerrim (-us)	acerrim	acerrim+e=acerrime
hirsutus	hirsutissim (-us)	hirsutissim	hirsutissim+e=hirsutissime

4. Second group. Adverbs with irregular grades of comparison. These adverbs are derived from the adjectives with irregular grades of comparison (cf. adjectives group (e) on p. 97 of this guide) and have correspondingly irregular grades of comparison. There are six such adverbs:

Adjective posit. degree	Adverb Positive	Comparative	Superlative
validus	valde	—	—
bonus	bene	melius	optime
malus	male	peius	pessime
magnus	magnopere	magis	maxime
parvus	parum	minus	minime
multus	multum	plus	plurimum

10. The syntax of the simple sentence.

§ 1. The structure of the simple sentence (general remarks).

In Latin each member of a sentence may occupy different places in this sentence depending on logical stress placed on this member. The simplest arrangement of words in the sentence is the following one:

Subject (first place); words describing subject (second place); words describing predicate (third place); predicate (last place).

Examples   (1, subject; 2, attribute; 3, object; 4, predicate):

1   2   3   4

Folia plantarum formam diversam habent.

1   2   4

Petioli foliorum Nelumbii longi sunt

§ 2. The members of the simple sentence.

1. In plant descriptions the subject is usually a noun. The subject always stands in Nominative case and, usually, in the first place in the sentence. The subject may be expressed by several nouns and may have the Plural number.

Examples (subject *italicized*):

Perigonium est organum externum floris.

Folia ex stipulis, petiolo et lamina constant.

Caulis, petioli et pedicelli pilosi sunt.

2. Predicate may be:
   a. Simple, if it is expressed by the verb only.
   b. Compound, if it is expressed by the verb joined with a noun, adjective, numeral, etc.

Examples (predicate *italicized*):

Group a. Flos hermaphroditus pistillum et stamina *habet*.
        Plantae aquaticae folia ampla *habent*.

Group b. Frutex *est planta arborescens*.
        Folia plantarum aliarum *sunt magna*.

3. The verb which is the part of a compound predicate is known as *copula*.

The copulas most frequently used are *est* and *sunt*, the forms of the verb *esse*=to be.

Examples (copula *italicized*):

Caulis *est* 30 cm. altus.

Pistillum et stamina *sunt* exserta.

Folia subtus *sunt* pilosa.

In the plant descriptions copula is usually omitted.

Examples:

Caulis 30 cm. altus.

Pistillum et stamina exserta.

Folia subutus pilosa.

The forms of the auxiliary verb *esse* are used also as a simple predicate.

Examples (predicate *italicized*):

Stipulae *adsunt*.

Indumentum *deest*.

Stolones *absunt*.

4. If an auxiliary verb is used as a simple predicate in negative form, the noun which is the object of negation, should stand in Nominative case.

Examples (object of negation *italicized*):

Planta acaulis *caulis* non habet.

Plantae cryptogamae *flores* non habent.

Forma inermis *aculei* non habet.

5. Besides the copula in a compound predicate, there are other parts of speech which describe or define subject. They are called Predicate Noun, Predicate Adjective, etc. In Latin these members of the sentence *must strictly agree in case and number with the subject*.

Examples:

Frutex est *planta arborescens*.

Frutices sunt *plantae arborescentes*.

Folium est *organum plantae*.

Folia sunt *organa plantae*.

6. The qualifying words (attributes) are the members of a sentence expressed by nouns, adjectives and participles. They answer to the questions: what? which? whose? how much/many?, and describe or define subject or object.

7. If logical stress do not lies on the qualifying word (attribute) the latter is usually placed *after* the word which it defines, but if the

stress lies on the qualifying word it is placed *before* the word which it defines.

Examples (qualifying words *italicized*):

a. Logical stress not on qualifying word.

Caulis *plantae*.	Squama *involucri*.	Hortus *botanicus*.
Frutex *scandens*.	Caulis *radicans*.	Species *secunda*.

b. Logical stress on the qualifying word.

Omnes plantae cryptogamae flores non habent.

Prima species est cum floribus rubris.

Tota planta cinereo-cana.

8. The attribute is called *agreed* when it agrees in case, gender and number with the word which it qualifies [defines]. Agreed attributes are expressed usually by adjectives.

Examples (agreed attributes *italicized*):

Dentes *calycini*.	Involucrum *florale*.
Planta *aquatica*.	Folia *ramealia*.

9. The attribute is called *non-agreed* when it does not agree in case, gender and number with the word which it qualifies [defines]. Non-agreed attributes are expressed usually by the noun in Genitive case.

Examples (non-agreed attribute *italicized*):

Dentes *calycis*.	Involucrum *floris*
Planta *aquarum*.	Folia *ramorum*.

10. Direct object is the thing which is immediately affected by the action expressed by a transitive verb. Direct object always stands in Accusative case.

Examples (object *italicized*):

Sepalum apice *mucronem* habet.	Papillae densae *capsulam* tegunt.

SECTION IV. THE BOTANICAL USAGE OF PARTS OF SPEECH.

1. The noun.

1. Nouns are widely used in plant descriptions. The names of plants and plant organs are expressed by nouns.
2. Nouns in the *Nominative case* are used as subjects in construction of sentences.
   Examples (nouns-subjects *italicized*):
   Planta annua; *fructus* globosus; *folia* viridia; *pedicelli* pilosi; *flores* lilacini; *ovarium* globosum; *caulis* striatus.
3. Nouns in the *Genitive case* may be used, instead of adjectives, for the qualifying other nouns or for indicating of appurtenance to something.
   Examples:
   a. Noun qualifying noun: Planta *aquarum;* academia *scientiarum;* familia *plantarum;* tentamen *florae Rossiae;* explicatio *tabulae;* index *florae Europae;* clavis *specierum;* index *herbariorum;* conspectus *systematis plantarum;* delectus *seminum, fructuum* et *sporarum.*
   b. Noun indicating appurtenance to something: Rami *inflorescentiae;* petala *corollae;* indusium *sori;* pinnae *folii;* bracteae *florum;* filamenta *staminum.*
4. If a noun is qualified with the help of another noun(s), it may be described in two ways:
   a. With the help of a noun only.
   b. With the help of a noun and preposition.
   Examples (qualifying nouns *italicized*):
   a. Noun qualifed with a noun only: squamae *involucri;* arista *glumae;* plantae *Manshuriae* et *Mongoliae;* flora *Sibiriae;* herbarium *academiae scientiarum;* habitus *plantae;* alae

seminis; pappus *achenii*; squamae *rhizomatis*.

b. Noun qualified with a noun and preposition: folia *cum dentibus*; caulis *cum ramis*; calyx *cum pilis*; semina *in fructu*; spadix *in spatha*. •

6. When such combinations of nouns, as previously discussed, are declined the noun that is qualified changes normally according to declension to which it belongs, but the qualifying noun remains *always* in Genitive case Singular or Plural.

Examples of declensions (qualifying noun *italicised*)

	Singular	Plural
N.	radius *umbellae*	radii *umbellae*
G.	radii *umbellae*	radiorum *umbellae*
D.	radio *umbellae*	radiis *umbellae*
Acc.	radium *umbellae*	radios *umbellae*
Abl.	radio *umbellae*	radiis *umbellae*

	Singular	Plural
N.	index *leguminosarum*	indices *leguminosarum*
G.	indicis *leguminosarum*	indicum *leguminosarum*
D.	indici *leguminosarum*	indicibus *leguminosarum*
Acc.	indicem *leguminosarum*	indices *leguminosarum*
Abl.	indice *leguminosarum*	indicibus *leguminosarum*

7. Nouns in the Dative case are used:

a. When describing affinity of one taxon to another with the help of adjectives: similis, -e; affinis, -e; propinquus, -a, -um.

Example (nouns in Dative case *italicised*):

Species haec *speciei Carex Komarovii* $\begin{Bmatrix} \text{similis} \\ \text{affinis} \\ \text{propinqua} \end{Bmatrix}$ est.

b. When describing the size of certain parts of a plant in relation with other parts of the same plant. This description is made with the help of adjectives: aequalis, -e; inaequalis, -e; aequilongus, -a -um; inaequilongus, -a, -um; subaequilongus, -a, -um.

In this case in Dative case is put the noun expressing the name
of plant's organ with which is compared another organ.

Example:

Internodia *foliis* | | aequalia
| | inaequalia
| | aequilonga
| | subaequilonga
| | inaequilonga

8. Nouns in the Accusative case are used to express the object in
   the sentences (cf. item 10 on p. 118 of this guide).

   Examples (object *italicized*):

   Folia apice *mucronem* habent; pili stellati *calycem* tegunt.

9. Nouns in the Ablative case are used to describe or indicate:

   a. The means by which something is done.

   Examples:

   Collum radicis *squamis* obtectum; caulis *pilis* longis vestitus;
   phylla involucri *appendicibus* instructa; folia *glandulis* cons-
   persa.

   b. Place.

   Examples:

   Basi; apice; margine.

   c. Time.

   Examples:

   Florendi tempore = tempore florifero; mense Julio; aetate; initio.

   d. Size of two organs compared with each other.

   Examples:

   Pedicellus *longitudine* calycis; petiolus *longitudine* laminae;
   fructus *magnitudine* Cerasi; sepala *latitudine* petalorum; ligula
   latitudine folii.

   e. Nouns stand also in Ablative case when they are associated with
      the preposition *cum*.

   Examples:

   Folia cum *stipulis*; planta cum *indumento*; achenium cum *rostro*;
   folia cum *vaginis*; calyx cum *glandulis*; petala floris cum

nectariis: collum radicis cum *squamis*.

2. The Adjective.

1. Adjectives are used to describe or define the nouns. Since nearly entire work of a taxonomist is to define or describe technically the characteristic traits of the plants and of their organs, it is natural that adjectives are the most widely used part of speech in botanical writings.

2. Adjectives are used to qualify nouns in botanical descriptions and in nomenclature as epithets of the taxa of specific and infraspecific rank.

3. The method of qualifying nouns with the help of adjectives is in all respects more convenient than that of qualifying with the help of nouns.

   Examples:

Qualifying with the help of noun(s):	Qualifying with the help of adjective(s):
Planta cum pilis et glandulis.	Planta pilosa et glandulosa.
Collum radicis sine squamis.	Collum radicis esquamatum.
Caulis cum ramis.	Caulis ramosus.

4. Adjectives in Latin stand, normally, *after* the nouns which they qualify and *must agree necessarily* with these nouns in gender, number and case.

   Examples:

Gender	Nominat. Sing.	Nominat. Plur.
Masculine	Fructus pilosus.	Fructus pilosi.
	Caulis strictus	Caules stricti.
	Dens calycis glandulosus.	Dentes calycis glandulosi.
Feminine	Planta sericea.	Plantae sericeae.
	Diagnosis plantae novae.	Diagnoses plantarum novarum
	Descriptio Latina ampla.	Descriptiones Latinae amplae.
Neuter	Folium tomentosum.	Folia tomentosa.
	Ovarium subglobosum.	Ovaria subglobosa.
	Semen minutum, alatum.	Semina minuta, alata.

5. For the correct agreement of nouns and adjectives it is essentially necessary, therefore, to know the gender of the noun(s), especially in second and third declensions where there are many nouns with irregular genus.

6. If several nouns are defined with one adjective, this adjective must stand in Plural and must agree in gender and case with the last noun.

Examples:

Folia, flores et fructus *magni*.

Folium, flos et fructus *magni*.

Bracteae, calyces et corollae *rubrae*.

Bractea, calyx et corolla *rubrae*.

Bractea, corolla et calyx *rubri*.

Each of the several nouns may be, however, defined separately. In this case the adjective must agree with the particular noun which it defines.

Examples:

Folium *magnum*, flos *magnus* et fructus *magnus*.

Folia *magna*, flores *magni* et fructus *magni*.

Bracteae *rubrae*, calyces *rubri* et corollae *rubrae*.

Bractea *rubra*, calyx *ruber* et corolla *rubra*.

7. Examples of usage of adjectives (adjectives *italicized*):

Diagnosis *Latina plena*; flos *albus plenus*; flores *albo-violacei*; folia *simplicia, rigida*; rami *floriferi, piliferi, aphylli*; segmenta *primaria ovata, secundaria lanceolata*; planta tota *sericeo-cana*; inflorescentia *terminalis, densa, pluriflora* [*multiflora*]; phylla perigonii *patentia, ovata, callosa*; culmus 8-9-*nodosus, glaber*, basi *geniculatus*; plantula *humilis, subtilis, delicatula, glabra*; bractea *ampla, cordata, spathiformis*; folia *ovata*, basi *subtruncata*, apice *acuminata*; caulis *teres striatus*, paulo *flexuosus*; rhizoma *longum, obliquum* vel *horizontale*; radix *crassa, lignosa, verticalis*.

8. When a noun associated with an adjective are both to be declined, *each member* of this combination is declined according to the

declension to which it belongs.

Examples:

a. Both noun and adjective of the first declension (feminine gender).

Singular	Plural
N.	planta pilosa	plantae pilosae
G.	plantae pilosae	plantarum pilosarum
D.	plantae pilosae	plantis pilosis
Acc.	plantam pilosam	plantas pilosas
Abl.	planta pilosa	plantis pilosis

b. Both noun and adjective of the second declension (masculine gender).

Singular	Plural
N.	nodus incrassatus	nodi incrassati
G.	nodi incrassati	nodorum incrassatorum
D.	nodo incrassato	nodis incrassatis
Acc.	nodum incrassatum	nodos incrassatos
Abl.	nodo incrassato	nodis incrassatis

c. Noun of the third declension, adjective of the first declension (feminine gender).

Singular	Plural
N.	radix longa	radices longae
G.	radicis longae	radicum longarum
D.	radici longae	radicibus longis
Acc.	radicem longam	radices longas
Abl.	radice longa	radicibus longis

d. Noun of the third declension, adjective of the second declension (neuter gender).

Singular	Plural
N.	nomen novum	nomina nova
G.	nominis novi	nominum novorum
D.	nomini novo	nominibus novis
Acc.	nomen novum	nomina nova
Abl.	nomine novo	nominibus novis

e. Noun of the fifth declension, adjective of the first declension (feminine gender).

	Singular	Plural
N.	species nova	species novae
G.	speciei novae	specierum novarum
D.	speciei novae	speciebus novis
Acc.	speciem novam	species novas
Abl.	specie nova	speciebus novis

f. Noun of the fourth declension and two adjectives, one of the third and the other of the second declension (masculine gender).

	Singular	Plural
N.	fructus viridis pilosus	frucutus virides pilosi
G.	fructus viridis pilosi	fructuum viridium pilosorum
D.	fructui viridi piloso	fructibus viridibus pilosis
Acc.	fructum viridem pilosum	fructus virides pilosos
Abl.	fructu viridi piloso	fructibus viridibus pilosis

9. The adjectives *inferus* and *superus* are used in botanical Latin in special meaning, as ovary [fruit] upper or lower: ovarium *superum/inferum*; fructus *superus/inferus*.

10. Comparative degree of the adjectives is utilized:

   a. In comparing of the qualities of several nouns.

   b. In qualifying of noun(s).

11. The comparison of qualities of several nouns may be expressed in two ways:

   (1) The noun *with which another one is compared* is placed in *Nominative* case with the word *quam*=than preceding this noun.

   Examples:

   Internodium quam *folium* longius; petioli quam *laminae* breviores; inflorescentiae quam *folia* longiores; bracteae quam *folia* minores; folia inferiora quam *folia superiora* majora.

   (2) The noun *with which another one is compared* is placed in *Ablative* case.

The word *quam* in the present instance is *not used*.

Examples:

Internodium *folio* longius; petioli *laminis* breviores; inflorescentiae *foliis* longiores; bracteae *foliis* minores; folia inferiora *foliis superioribus* majora; petala *calyce* longiora; dentes calycis *tubo* breviores.

12. The expressions, such as "pubescence more dense", "leaves more broad", etc. are translated in Latin with the help of adjectives in the *comparative* degree. The above given sentences will read in Latin as: *pubescentia densior* and *folia latiora*.

13. The same type of expressions as previously mentioned may be translated into Latin also with the help of words *magis* or *plus=* more.

Examples:

Pubescence more dense=pubescentia magis densa=pubescentia plus densa. Leaves more broad=folia magis lata=folia plus lata.

14. The expressions like "pubescence less dense" or "leaves less broad", etc. are translated with the help of word *minus=*less.

Examples:

Pubescence less dense=pubescentia minus densa.

Leaves less broad=folia minus lata.

15. Superlative degree is utilized in expressions like: "bright yellow" or "bright red", "intensely green", "completely glabrous", "completely entire", etc.

Examples:

bright yellow=flavissimus, -a, -um; bright red=ruberrimus, -a, -um; intensely green=viridissimus, -a, -um; completely glabrous =glaberrimus, -a, -um; completely entire=integerrimus, -a, -um. In such adjectives, describing color, as: *cyaneus, violaceus, brunneus*, etc. superlative degree is formed with the help of the word *intense=*intensely. Hence bright blue will be *intense cyaneus,* bright violet *intense violaceus,* etc.

16. Superlative degree of the adjectives is used also for describing of qualities of nouns which are developed to the highest extent, e.g., leaves very narrow, pubescence very dense, etc. (cf. also item 15 on p. 126 and examples).

Examples:
Leaves very [most] narrow=folia *angustissima*.
Pubescence very [most] dense=pubescentia *densissima*.
Stem very [most] prickly=caulis *spinosissimus*.

17. Examples of usage of degrees of comparison of adjectives:
Folia lineari-lanceolata *angustissima*; planta cum pubescentia *densissima*; plantula parva, *humillima*; calyx *glaberrimus*; corolla *flavissima*; fructus *ruberrimi*; folia margine *integerrima*; pedicelli *magis* [plus] *villosi*; folia mediana quam inferiora *pilosior*; pedicelli quam calyx *minus glandulosi*; folia supra *minus tomentosa*.

3. Adjectives derived from nouns and formation of compound adjectives.

1. The adjectival botanical terms which describe characters of plant organs are often derived from the nouns. The formation of these adjectives is made with the help of certain suffixes.

Examples:

Noun	Adjective
acus, -us=sharp point	acutus, -a, -um =acute
sinus, -us=sinus	sinuatus, -a, -um=sinuate
pilus, -i=hair	pilosus, -a, -um=pilose
axilla, -ae=axil	axillaris, -is, -e=axillary
ochrea, -ae=ocrea	ochreatus, -a, -um=ocreate

2. In some cases it is even possible to make several adjectives of different meaning and different declensions from one noun.

Examples:

Noun	Adjective
petiolus, -i=petiole	petiolaris, -is, -e =pertaining to petiole
	petiolatus, -a, -um=having petiole, petioled
cilium, -ii=cilium	ciliaris, -is, -e=of cilia
	ciliatus, -a, -um=ciliate, furnished with cilia

3. Compound adjectival botanical terms are formed by combination in one word of:

a. Two nouns joined together with connecting vowel i.
Examples:
squama+forma=squamiformis, -is, -e
ren, -nis+forma=reniformis, -is, -e
cor, -dis+forma=cordiformis, -is, -e

b. Adjective and noun.
Examples:
rigidus+folium=rigidifolius, -a, -um
tenuis+caulis=tenuicaulis, -is, -e

c. Numeral or adjective showing quantity and noun.
Examples:
tri+foliolus=trifoliolatus, -a, -um
multi+flores=multiflorus, -a, -um
bi+color=bicolor, -oris.

d. Two adjectives, chiefly those describing color or form. The components of the compound adjective are divided by hyphen. There are two subgroups of these adjectives: (1) The first component of the compound adjective is the adjective of the first or second declension. In this case the components of compound adjective are joined together with the connecting letter o.
Examples:
albus+violaceus=albo-violaceus, -a, -um
flavus+viridis=flavo-viridis, -is, -e
oblongus+ellipticus=oblongo-ellipticus, -a, -um
crispatus+villosus=crispato-villosus, -a, -um

(2) The first component of the compound adjective is the adjective of the third declension. In this case the components of the compound adjective are joined together with the connecting letter i. Examples:
triangularis+ovatus=triangulari-ovatus, -a, -um
viridis+fuscus=viridi-fuscus, -a, -um

ciliaris + pilosus = ciliari-pilosus, -a, -um

longitudinalis + striatus = longitudinali-striatus, -a, -um

e. An adjective and a prefix; the most common prefixes used for the formation of compound adjectives are: e-(ex-), a-, in- = not, without; and sub- = nearly, almost.

Examples:

e + dentatus = edentatus, -a, -um

ex + alatus = exalatus, -a, -um

a + phyllus = aphyllus, -a, -um

a + caulis = acaulis, -is, -e

in + articulatus = inarticulatus, -a, -um

in + dehiscens = indehiscens, -entis

sub + acutus = subacutus, -a, -um

sub + globosus = subglobosus, -a, -um

f. An adjective with participle or numeral.

Examples:

tri + sectus = trisectus, -a, -um

bi + fidus = bifidus, -a, -um

g. A noun and suffixes -fer or ger = bearing.

Examples:

flos, -ris + fer(ger) = florifer (floriger), -a, -um

pilus, -i + fer(ger) = pilifer (piliger), -a, -um

ramus, -i + fer(ger) = ramifer (ramiger), -a, -um

h. A noun and suffix -ideus = similar to.

Examples:

petalum + ideus = petaloideus, -a, -um

botrys + ideus = botryoideus, -a, -um

bulbus + ideus = bulboideus, -a, -um

Note: In the g. group the components of the compound adjective are joined together with the help of the connecting letter *i*, while in h. group they are joined together with the help of the connecting letter *o*.

4. The verb.

1. The use of verbs is limited in plant descriptions. More often for this purpose some of their forms are utilized.

2. Personal forms of the auxiliary verb *esse* = to be, are seldom employed by taxonomists, but they are basic for the forming of following four words used in plant descriptions:

   adest = he, she, it is present   abest or deest = he, she, it is absent
   adsunt = they are present   absunt or desunt = they are absent.

3. Below are listed some of the forms of Latin verbs which are more often used in taxonomical papers:

   addo = I add to (from addo, -ĕre)
   conservatur = it is preserved (from conservo, -are)
   corrigo = I correct; corrigit = he, she corrected (from corrigo, -ĕre)
   crescit = it grows (from cresco, -ĕre)
   delineavit = he, she drew (from delineo, -are)
   determinavit = he, she determined (from determino, -are)
   differt = it differs (from differo, -ĕre)
   discrepat = it differs (from discrepo, -are).
   distinguitur = it is distinct (from distinguo, -ĕre)
   elaboravit = he, she, worked out (from elaboro, -are)
   emendavit = he, she amended (from emendo, -are)
   floret = it flowers [blooms] (from florēo, -ēre)
   habet = it has; habent = they have (from habēo, -ēre)
   habitat = it inhabits [lives] (from habito, -are)
   legit = he, she gathered [collected] from lego, -ĕre)
   nascet = it is born [is used in the expression: sponte nascet = occurs wild] (from nasco, -ĕre)
   nominavi = I named (from nomino, -are)
   observavit = he, she observed (from observo, -are)
   occurrit = it occurs (from occurro, -ĕre)
   tegit = it covers; tegunt = they cover (from tego, -ĕre)
   vidi = I saw (from vidēo, -ēre).

4. Examples of botanical usage of verb forms (verb forms itali-
    cized):

Ad diagnosin *addo* descriptio florum; typus in herbario academiae scientiarum *conservatur*; *addenda* et *corrigenda*; *crescit* in
paludibus Europae mediae; auctor *delineavit; determinavit* Komarov;
differt a typo floribus majoribus; floribus majoribus a typo *discrepat;*
species haec statura majore et toto habitu statim *distinguitur;*
elaboravit T. Nakai; Sect. *Microdiscus* Fed. *emendavit* Baranov;
habitat in silvis; *floret* mense Junio; varietas haec folia minora
habet; floret in Maio; auctor *legit;* in pratis Chinae borealis *sponte
nascet;* varietatem hanc in honorem botanici F. Schmidtii *nominavi;*
in silvis Sibiriae sparse *occurrit;* tomentum densum fola *tegit;*
squamae densae collum radicis *tegunt;* plantam parvam *vidi;* flores
cleistogamos non *vidi;* flores cleistogami non *observati;* folia radicalia
desunt; nectarium *adest;* dentes in margine folii *absunt;* pedicellus,
receptaculum, stamina et pistillum *sunt* partes floris; carina, vexillum et alae *sunt* petala floris Leguminosarum.

5. When describing the differences among several taxa with the
    help of verb forms, as *differt, discrepat,* etc., the nouns and
    adjectives expressing the characters of difference are put in
    Ablative case.

Example:

Species A a specie B *caule hirsuto, foliis denatatis, floribus
majoribus,* etc., differt [or discrepat].

5. The participle.

1. Participles are often used in botanical nomenclature as epithets
    of species and/or infraspecific taxa, and as adjectival botanical
    terms in plant descriptions for qualifying nouns.

2. Below are listed the participles most commonly used in descrip-
    tive botany:

-fidus, -a, -um = incised

glabrescens, -entis = glabres-
                        cent

glaucescens, -entis = glauces-
                        cent

instructus, -a, -um = furnished

nutans, -antis = nodding

obtectus, -a, -um = covered, clothed

-partitus, -a, -um = parted

pubescens, -entis = pubescent

radicans, -antis = rooting

rufescens, -entis = becoming rufous

scandens, -entis = climbing

-sectus, -a, -um = sected [divided]

vestitus, -a, -um = covered, clothed.

3. Examples of usage of participles (participles are *italicized*):
Planta initio pilosa aetate *glabrescens*; folia *glaucescentia*; squamae involucri appendicibus *instructae*; collum radicis squamis *obtectum*; caulis *pubescens*; frutex *scandens*; herba perennis pilis albis dense *vestita*; folium *pinnatifidum*; bractea *plamati-secta*; stylus *tripartitus*; planta *florens*; planta *fructificans*; caulis *ascendens*; corolla *patens*; caulis in nodos *radicans*; flores *nutantes*.

4. Likewise adjectives, the participles usually stand *after* the noun which they qualify and must necessarily agree with this noun in number, case and gender.

6. The preposition.

1. Prepositions are the words connecting elements of the sentence and expressing relations of these elements to each other.

2. As is known, nouns associated with prepositions must be put either in Accusative or in Ablative case. The rules relating thereto are explained on p. 104,105 of this guide. More frequently than other, are used prepositions cum (with a noun in Ablative case) and in (with a noun in Accusative case when preposition indicates motion towards, or a noun in Ablative case when preposition indicates fixed position of something).

3. Examples of usage of prepositions (prepositions *italicized*):
In apice; *in* apicem; explorationes plantarum *in* Brasilia; palea apice *in* aristam longam attenuata; contributio *ad* cognitionem florae Manshuricae; folia *sine* vaginis; calyx *cum* nectariis;

notulae *de* flora Manshuriae et Mongoliae; supplementum *ad* indicem specierum novarum; axis inflorescentiae *pro* norma *cum* pilis; addenda *ad* catalogum plantarum; sori *secus* nervos dispositi; petiolus basin *versus* alatus; fructus globosi *cum* pedunculis longis; lamina *per* marginem crenata; folia *in* margine crenata; folia *ad* basin dilatata; tubus corollae *in* media parte constrictus; *a* typo differt; *ex* axillis; involucri phylla *sine* appendicibus; *pro* specie; *pro* parte; *sub* anthesin; *sub* folio; *sub* folium; *ante* anthesin; *circum* axem; *extra* vaginam; *intra* vaginam; *post* anthesin; *supra* folium; *inter* venas.

7. The conjunction.

1. Conjunctions are used in plant descriptions for joining of different members of the sentence.

2. As is known, conjunctions *-que* and *-ve* are joined to the end of the words and conjunction *autem* is never put in the beginning of the sentence (cf. p. 106 of this guide).

3. Examples of usage of conjunctions (conjunctions *italicized*):

   Lamina petioli*que* villosi; lamina *et* petioli villosi; caulis ad nodos plus minus*ve* incrassatus; folia, petioli *atque* pedicelli tomentosi; folia longa *ac* lata; flores rubri *vel* purpureo-rubri; diagnosis *sive* descriptio plantae; caulis basi subglaber, *sed* in parte superiore plus minus pilosus; caulis basi subglaber in parte superiore *autem* pilosus; semina *aut* nigra *aut* atro-brunnea; folia *ut* in typo; flores cleistogami clausi chasmogami *tamen* patentes.

8. The numeral.

1. Numerals are used very often in plant descriptions for showing of size of plants and of number and dimensions of their organs. Cardinal numbers are usually not spelled, but shown by figures. Examples:

   Caulis 43 cm. altus; folia 30 mm. longa, 8 mm. lata; inflorescentia 12-35-flora; fructus 3-carpellatus; culmus 8-9-nodosus, etc.

2. If the number is small it is sometimes spelled, e.g., flos quinque-petalus; folia trinervia; fructus quadriloculatus; inflorescentia triflora.

3. Sentences describing size and number may be constructed in different ways:

(a) With the help of preposition *cum*.

Examples:

Planta *cum* altitudine 24 cm. Folia *cum* longitudine 5 cm. et latitudine 3 cm.

Caulis *cum* crassitudine   Inflorescentia *cum* 12-35 floribus. 0,5 cm.

(b) Without preposition cum, only by placing nouns *altitudo, -inis; crassitudo, inis,* etc. in the Ablative case.

Examples:

Planta *altitudine* 24 cm.   Folia *longitudine* 5 cm. et *latitu-* Caulis *crassitudine* 0,5 cm.   *dine* 3 cm.

(c) With the same nouns, but placed in the Nominative case and in the beginning of the sentence.

Examples:

Altitudo plantae 24 cm.   *Longitudo* foliorum 5 cm. et *Crassitudo* caulis 0,5 cm.   *latitudo* 3 cm.

(d) With the help of adjectives: altus, -a, -um; crassus, -a, -um; longus, -a, -um; latus, -a, -um; -florus, -a, -um.

Examples

Planta 24 cm. *alta.*   Folia 5 cm. *longa,* 3 cm. *lata.* Caulis 0,5 cm. *crassus.*   Inflorescentia 12-35-*flora.*

The last way (paragraph d.) is the most concise, convenient and the most used in plant descriptions.

4. When the thickness of some plant organs, or diameter of the flowers are described, the word *diametrus, -i* is usually used which is put in the Ablative case. With regard to stems and branches this

noun is used instead of adjective crassus, -a, -um.

Examples:

Fructus *diametro* 1,3 cm.   Flores sub anthesin 2,5 cm.

                                                       diametro.

Pedicelli 0,2 cm. *diametro.*   Petiolus 1,5 mm. *diametro.*

[Pedicelli 0,2 cm. crassi]   [Petiolus 1,5 mm. crassus]

Caulis 0,8 cm. *diametro.*   Ovarium ca. 1 cm. *diametro.*

[Caulis 0,8 cm. crassus]

Calathium 2,5 cm. *diametro*   Umbelluli ca. 1,4 cm. *diametro.*

5. When the number of plant organs is described in some cases this is done with the help of the word *numero* (Ablative form of the noun numerus, -i = number).

    Examples:

    Flores *numero* 12-35   Petala floris *numero* 4.

    Fructus *numero* 3.   Stamina *numero* 5.

6. Cardinal Roman numerals were always used by early authors to show the year of publication of their papers. Three ways to write the years of publication are known: 1. MDCCLX; 2. M DCC LX; 3, M.DCC.LX which in all cases means the same, i.e. the year 1760.

    Examples of reading of the years:

    M DCC LX = 1760   M DCCC LIX = 1859

    M DCC LXXV = 1775   M DCCC LXII = 1862

    M DCC LXXXII = 1782   M DCCCC LI = M CM LI = 1951

    M DCCC XXVII := 1827   M DCCCC LX = M CM LX = 1960

    M DCC VI = 1706   M DCCCC XXVII = M CM XXVII

                                                                  = 1927

7. Ordinal numesral, up to ten, are usually spelled in plant descriptions.

    Example:

    Species haec affinis a speciebus A, B et C, sed a *prima* differt caule erecto, a *secunda* foliis majoribus et a *tertia* floribus roseis.

8. Ordinal numerals are also used for numbering of parts of books, volumes, plates, etc. In this case they are *not* spelled.

Examples:

Pars I=part 1-st; fasciculus IV=fascicle 4-th; supplementum XII=supplement 12-th; tomus V=volume 5-th; tab. CXXI= plate 121-st.

9. The numerical prefixes (cf. p. 109 of this guide) are used in following way:

(a) The prefix *sesqui* is used by itself. This word being prefixed to the word expressing measure shows that a plant or its organ exceeds another one or certain measure by one half.

Examples:

Frutex sesquimetralis; internodia petiolis sesquilongiora; bracteae pedicellis sesquibreviores.

(b) Other prefixes are used for forming numerical adjectives and adverbs.

Examples of numerical adjectives:

bipinnatus, -a, -um = duplicately pinnate; bifidus, -a, -um=fidb; biternatus, -a, -um=duplicately ternate; duplicatus, -a, -um= duplicated.

Tripinnatus, -a, -um=triplicately pinnate; trifidus, -a, -um= trifid; triternatus, -a, -um=triplicately ternate, etc.

Quadripinnatus, -a, -um=quadruplicately pinnate; quadrifidus, -a, -um=quadrifid; multifidus, -a, -um=multifid; multisectus, -a, -um=multisected; semiovatus, -a, -um=semiovate, etc.; sesquilongior, -ius=one and a half times longer; sesquimetralis, -is, -e=1.5 m. long, broad or high.

Examples of numerical adverbs:

duplo=twice as great; triplo=thrice as great; quadruplo=four times as great, etc.

Examples of multiplicative adjective:

simplex, -icis = simple, single, once: duplex, -icis=double; triplex, -icis=triple; multiplex, -icis=manifold.

10. Examples of usage of numerals (numerals *italicized*):

Flores *terni*; calyx *quadrifidus*; corolla tubulosa cum limbo *quin-
quelobo*; frondes *triplicato-ternatae*; caulis pilis *bifurcatis*
obtectus; inflorescentia *multiflora*; folia *multisecta* vel *multi-
fida*; petiolus lamina *duplo* longior; sori *bini*; planta *sesquimetralis*; folium pedicello *sesquilongius*; stipulae *semicordatae*.

9. The pronoun.

1. Pronouns are comparatively rarely used in plant descriptions.

2. Demonstrative pronouns which are sometimes utilized by taxonomists are forms of pronoun *hic, haec, hoc*=this, in the following
expressions:

Species *haec* caule prostrato et floribus minoribus distincta est,
which means this species is distinguished by prostrate stem
and smaller flowers.

Speciem *hanc* in honorem Pallasii botanici clarissimi nominavi,
which means I named this species in honor of the renown
botanist Pallas.

3. Since this pronoun (*hic, haec, hoc*) has generic forms and is
declinable it necessarily must agree with the noun with which it
is associated in case, gender and number. In both examples given
above this pronoun agrees with the word *species* and is in the
first example in Nominative case Singular and in second example
in Accusative case Singular.

4. The personal pronoun *mihi* the Dative Singular form of pronoun
ego=I is also sometimes used in plant descriptions. It is used
in the expression, this species (variety, genus, etc.) is described
by me (literally to me) which reads in Latin in following way:
species/varietas haec mihi descripta est; genus hoc mihi des-
criptum est. The pronoun *nobis* (literally to us) the Dative form
of pronoun *nos*=we, is sometimes utilized as well as the word
mihi to designate autority of plant taxa.

Example:

Adenophora mongolica *mihi* (Singular, used in case of one author).

Adenophora mongolica *nobis* (Plural, used in case of several authors).

5. The feminine form of possessive pronoun *noster, -a, -um*=our is used in expressions, *tabula nostra*=our plate or *figura nostra*= our figure to indicate plates or figures attached by a given author to his paper.

6. The Ablative form of relative pronoun *qui, quae, quod*=which is used in the expression, *ex qua differt* (feminine) or *ex quo differt* (masc./neuter).

Examples:

Species prima proxima est a specie secunda *ex qua* differt floribus magnis, etc.

Genus A similis est a genere B *ex quo* differt staminibus connatis, etc.

7. The Genitive form of the same relative pronoun *qui, quae, quod* is utilized to express one portion of a whole which is compared with the remaining part.

Example:

Inflorescentia cum ramis tribus *quorum* medius longissimus est.

Flores cum bracteis *quarum* imae maximae sunt.

10. The adverb.

1. Adverbs are words used in the plant descriptions to modify a verb or adjective by describing time, place, manner, degree, etc.

Examples of usage of adverbs (adverbs *italicized*):

(Time) Folia *nunquam* dentata; rami inflorescentiae *semper* verticillati; planta *primum* dense pilosa *demum* (or *mox*) glabrescens.

(Place) Folia *subtus* tomentosa; caulis *sursum* glanduloso-pilosus; bracteae *supra* nitentes.

(Manner) Rhizoma *longe, horizontaliter* repens; involucri phylla *seriatim* dispositi: folia *pinnatim*, *regulariter* incisa.

(Degree) Planta *dense* villosa; caulis *sparsim* pilosus; inflorescentia *paulo* nutans; frutex *valde* ramosus.

Folia *semper* apicem versus *sensim* angustata; planta viridis *interdum* glaucescens; petala apice *abrupte* (or *subito*) dilatata; caulis *praecipue* in parte superiore *sparsim* glandulosus; tubus corollae *aliquanto* infra medium *leviter* constrictus; foliola apice *non nunquam* *vix* emarginata; stipulae *rariter* incisae: pedunculus *plerumque* glaber *sat* glandulosus.

11. Sentences expressing negation.

1. For expressing negation in Latin are used the following words:
    Adverbs, *non* = not, no; *nunquam* = never;
    Conjunctions, *nec*, *neque* = and not;
    Pronoun, nullus, -a, -um = none. This is the only of negatives which must necessarily agree grammatically with the word with which it is associated.
    Preposition, *sine* = without.

2. Negative sentences in which nouns are defined with the help of other noun(s) have negation expressed by preposition *sine* followed by noun(s) in Ablative case. But if the noun(s) is/are defined by adjective(s), the negation is expressed by the adverb *non*.
    Examples:

Noun(s) defined with the help of noun(s)	Noun(s) defined with the help of adjective (s)
Folia *sine* pilis	Folia *non* pilosa
Flores *sine* bracteis	Flores *non* bracteati
Caulis *sine* ramis	Caulis *non* ramosus.

3. Examples of usage of negatives (negatives *italicized*):
    Caulis *non* striatus; rami *non* verticillati; calyx *non* pilosus *nec* glandulosus; planta *non* alta *nec* robusta; flores *nunquam* nutantes; rhizoma *nunquam* repens; folia subtus *nunquam* tomentosa;

planta cum pubescentia *nulla*; folia radicalia *nulla*; vaginae foliorum *nullae*: pili glandulosi *nulli*.

4. It should be clearly understood that in Latin only one negation in each sentence is possible. Two negations joined together give affirmation.

Examples:

Nunquam=never, non nunquam=sometimes, nunquam non=always.

Nullus=none, non nullus=some.

Nec=and not, non nec=and also.

Examples of usage of combinations of negatives:

Caulis subglaber vel *non nunquam* pubescens; bracteae ellipticae vel *non nunquam* ovatae; descriptiones plantarum *nonnullium* Americae; flora regionis Ussuriensis *nec non* Manshuricae.

5. All sentences in plant descriptions showing absence of some characters are constructed with the help of the following words: the adjective *ignotus, -a, -um*=unknown: the expression *non vidi*=I have not seen; and the verbs *deest*=it is absent or *desunt*=they are absent.

Of these only adjective *ignotus* must necessarily agree grammatically with the noun with which it is associated.

6. The adjective *ignotus* is used in two senses:

(1) In the sense of "absent" likewise the words *non vidi* and *deest/desunt*. Instead of all these words question mark or ellipsis may be used as well.

Examples:

Radix *ignota*=radix *deest*=radix *non vidi*=radix?=radix...= root unknown, i.e. absent.

Flores *ignoti*=flores *desunt*=flores *non vidi*=flores?=flores...= flowers unknown, i.e. absent.

(2) In the sense of "unknown". This happens when are cited specimens on the labels of which certain data are lacking.

Examples:

Locus *ignotus*=locus?=locus...=locality (where the specimen was collected) is unknown.

Datum *ignotum*=datum? =datum...=date (of collecting) is unknown.

Collector *ignotus*=collector? =collector...or legit...=collector is unknown.

Examples of citation of incomplete labels:

Asia Media, locus ignotus, ad ripas fluvii, 2.VII. 1954, legit M. Popov.

Asia Media, locus? [locus...], ad ripas fluvii, 2.VII. 1954, legit M. Popov.

Mongolia Interior, prope Hailar, datum ignotum [or datum? or datum...] leg. C. C. Wang no. 610.

China, prov. Liaoning, prope Shenyang (Mukden olim), 20.VI. 1950, collector ignotus [or collector? or collector...]

China, prov. Jeho, 13. IX. 1952, collector ignotus [or collector? or collector...] no. 5036.

China, prov. Shan-tung, sine loco speciali, *Y. Nagai no. 55.* 1914.

Japonia, sine loco speciali, 3. VII. 1886, leg. Mayr s.n.

China, Taiwan, Mt. Arisan, *Kanehira & Sasaki s.n.*

Note: The expression *sine loco speciali* means without further locality. It is used when the exact locality where the specimen was collected is not recorded in the label. The abbreviation *s.n.* or *sine num.* means without number and is used when collector has not numbered given specimen.

12. Variation and abbreviation of sentences.

1. One of the basic properties of the Latin language is the tendency to express one's thoughts as briefly as possible.

2. The richness of Latin in terms and expressions together with possibility of abbreviation of sentences by omitting certain words,

enables the expression of one and the same thought in various manners with regard to form of presentation and shortness. In the following examples the shortest form is the last one.

Examples:

		1. Caulis in parte superiore ramos ferens (or ramos emittens)
Stem in the upper part branched		2. Caulis in parte superiore ramis instructus.
		3. Caulis in parte superiore cum ramis.
		4. Caulis parte superiore cum ramis.
		5. Caulis parte superiore ramosus.

		1. Folium in margine cum dentibus.
Leaf dentate along the margin		2. Folium per marginem dentatum.
		3. Folium margine cum dentibus.
		4. Folium in margine dentatum.
		5. Folium margine dentatum.

3. The abbreviation of sentences is made by omitting certain words, the absence of which does not detract from the intelligibility of the context. The words which may be omitted are : verbs, prepositions, pronouns and some nouns.

Examples of abbreviation of sentences:

N o n - a b b r e v i a t e d	A b b r e v i a t e d
Planta robusta *est*	Planta robusta.
Planta perennis	Perennis.
Caulis *est* striatus	Caulis striatus.
Caulis *in* apice pilosus	Caulis apice pilosus.
Folia *cum* margine integro	Folia margine integra.
Calyx *cum* tubo angusto	Calyx tubo angusto.

SECT. V. GLOSSARY OF SELECTED STANDARD TERMS AND ABBREVIATIONS USED IN TAXONOMICAL WORK.

Addenda	Additions
ad verg.=ad vergens	tending towards such and such a taxon or to such and such a taxon approaching.
Affinis	related to such and such a taxon.
An sp. (gen., var., etc.) nov.?	Perhaps new species (genus, variety etc.)
A priori	Beforehand, without proving.
Auct. =auctorum	of the authors.
Auct. fl.=auctorum florae	of the authors of the flora (such and such).
ca.=circa	about, approximately.
Cf., cfr.=confer, conferatur	compare
Comb. nov.=combinatio nova	New nomenclatural combination
Corrigenda	corrections
Descr. ampla=descriptio ampla	detailed description
Descr. emend.=descriptio emendata	corrected description
Diagn. brev. = diagnosis brevis	brief diagnosis
E. grege	out of (such and such) a group
Ex parte	partly
E. g.=exempli gratia	for example
Errore	erroneously
Ex majore parte	for the greater part
Excl. =excluso	excluding
Fide	according to
Floret	it flowers, e.g., *floret in Maio*=it flowers in May
Florende tempore	during flowering time
Grad. nov.=gradus novus	new rank; this expression is used when only rank, but not the name of the taxon is changed
Hort.=hortorum or hortulanorum	of gardens or of gardeners

Ibidem	in the same place; the same; ibidem
Id.=idem	the same
In oculo armato	under magnifying glass
In oculo nudo	by naked eye
In sched.=in schedis	on the label
In sicco	in dry condition
In textu	in the text
In vivo	in living or in fresh condition
Ined.=ineditus. -a, -um	not published
L.c.=loco citato	in cited paper
Leg.=legit	collected
Nomen conservandum	conserved name
Nomen illegitimum	illegitimate name
Nomen novum	new name
Nomen nudum	name published without diagnosis
Nomen rejiciendum	rejected name
Nomen subnudum [seminudum]	name published with partial observing of the rules of effective publication
Oculo nudo	same as *in oculo nudo*
P.=pagina	page
Partim	partly
Per lentem	under magnifier
pro	as, e.g., *pro specie*=as a species; in the rank of species
Pro parte	partly
Prope	near, not far from
Sensu amplo	in broad sense
Sensu lato	same as preceding
Sensu stricto	in narrow sense
Sphalmate	erroneously
Stat. nov.=status novus	new taxonomic status; new rank
Sub anthesin	in flowering time
Sub lente, Sub microscopio	same as *per lentem* or *in oculo armato*
Tabula	plate
Tabula nostra	our plate, i.e. the plate attached to the given paper
Teste	in accordance with (such and such author); sometimes is used also instead of *deter-*

	minavit=determined by
V. =vide	see
Via ferrea	railway; hence *statio viae ferreae*=railway station
Vol. =volumen	volume.

146

LITERATURE

Greenough, J. B. et al., editors. Allen and Greenough's New Latin Grammar. Ginn & Co. Boston, etc. 1931.

Knipovich, M. F. et al. Latinskii yazyk [Latin language for secondary medical schools]. Moscow. 1949. (in Russian).

Fedorov, An. et Kirpicznikov, M. Vademecum methodi systematis plantarum vascularium. Fasc. I. Abbreviationes, designationes institutae, nomina geographica. Moscow-Leningrad. 1954.

Stearn, W. T. Botanical Latin. London. 1966.

Zabinkova. N. et Kirpicznikov, M. Vademecum methodi systematis plantarum vascularium. Fasc. II. Lexicon latino-rossicum pro botanicis. Moscow-Leningrad. 1957.